Control of Scale and Corrosion in Building Water Systems

Russell W. Lane, P.E.
Water Treatment Consultant

McGraw-Hill, Inc.
New York San Francisco Washington, D.C. Auckland Bogotá
Caracas Lisbon London Madrid Mexico City Milan
Montreal New Delhi San Juan Singapore
Sydney Tokyo Toronto

Library of Congress Cataloging-in-Publication Data

Lane, R. W.
    Control of scale and corrosion in building water systems / Russell W. Lane.
      p.   cm.
    Includes bibliographical references and index.
    ISBN 0-07-036217-3
    1. Water-pipes—Maintenance and repair. 2. Water-pipes-Corrosion.
    3. Pipes, Deposits in. I. Title.
    TD491.L36   1993
    696'.1—dc20   92-35156
                                                                                             CIP

Copyright © 1993 by McGraw-Hill, Inc. All rights reserved. Printed in the United States of America. Except as permitted under the United States Copyright Act of 1976, no part of this publication may be reproduced or distributed in any form or by any means, or stored in a database or retrieval system, without the prior written permission of the publisher.

1 2 3 4 5 6 7 8 9 0   DOC/DOC   9 9 8 7 6 5 4 3

0-07-036217-3

The editors for this book were Robert W. Hauserman and Laura Givner, and the production supervisor was Pamela A. Pelton. It was set in Century Schoolbook by McGraw-Hill's Professional Book Group composition unit.

Printed and bound by R. R. Donnelley & Sons Company.

Information contained in this work has been obtained by McGraw-Hill, Inc., from sources believed to be reliable. However, neither McGraw-Hill nor its authors guarantees the accuracy or completeness of any information published herein and neither McGraw-Hill nor its authors shall be responsible for any errors, omissions, or damages arising out of use of this information. This work is published with the understanding that McGraw-Hill and its authors are supplying information but are not attempting to render engineering or other professional services. If such services are required, the assistance of an appropriate professional should be sought.

Contents

Preface   vii

Chapter 1. Introduction   1

    Water and Its Properties   5
    The Composition of Waters   5
    Expression of Concentrations of Water Constituents   7
    Most Common Causes of Corrosion and Scale Problems   8
    References   9

Chapter 2. Corrosion   11

    Corrosion—An Electrochemical Process   11
    Types of Corrosion   16
    Anodic, Cathodic, and Mixed Inhibitors   19
    Water Constituents and Properties—Important in the Corrosion Process   20
    Corrosion Testing   21
    Corrosion of Galvanized Steel in Building Water Systems   25
    Corrosion of Galvanized Piping by Water   25
    Design of Domestic Hot Water Systems   30
    Cathodic Protection   30
    Electric Grounding   31
    Corrosion of Copper in Building Water Systems   31
    Deficiencies of Copper as a Corrosion-Resistant Metal   33
    Pitting of Copper   35
    Water Treatment Methods for Inhibiting Copper Corrosion   37
    Copper Concentrations   38
    Corrosion in Tanks   39
    Corrosion of Brass   39
    Case Histories   40
    References   40

Chapter 3. Potable Water — 43

Criteria for Potable Water — 43
Lead Contamination — 45
Chilled Potable Water — 46
Chlorination — 47
POU Treatment — 48
Bottled Water — 49
Corrosion Inhibitor Treatment of Potable Water — 49
Domestic Hot Water — 50
Corrosion Testing — 52
Corrosion of Copper and Its Role in Corrosion of Galvanized Steel — 53
Indexes for Estimating Corrosive and Scale Tendencies of a Water — 53
Metallic Contamination of Drinking Water — 60
References — 67

Chapter 4. Steam and Hot Water Heating Systems — 69

Regulations — 70
General Steam Power Plant Operations — 70
Boiler Water Treatment — 72
Proper Water and Steam Sampling — 84
Chemical Feed Systems — 86
Continuous Blowdown — 87
Discussion of Boiler Water Test Limits — 90
Demineralization or Dealkalization — 92
Treatment for Condensate Return Systems — 93
Operation and Proper Maintenance — 95
Steam Purity and Its Control — 103
Maintaining Efficient Plant Operation — 105
Operational Techniques for Improving Internal Boiler Conditions — 106
Boiling Out New Boilers — 106
Proper Lay-up of Boilers When Out of Service — 107
References — 109

Chapter 5. Open Recirculating Cooling Water Systems and Treatment — 111

Free, or Ambient, Cooling — 112
Cooling Towers — 113
Water Treatment and Water-Testing Equipment — 118
Objectives of Cooling Water Treatment — 122
Cathodic Protection — 143
Legionnaire's Disease — 143
Initial Conditioning of Cooling Water Equipment — 145
Cooling System Lay-up — 146
References — 147

Chapter 6. Closed Hot and Chilled Water Systems and Treatment   149

    Definition of a Closed System   149
    Corrosion and Its Control   150
    Fouling and Its Control   151
    Source of Leakage   154
    Thermal Energy Systems   155
    Free Cooling   155
    Glycol System   156
    Hot Water Closed Systems   158
    Chemical Treatment   159
    Low-Pressure Steam or Hot Water Boilers   162
    Corrosion Testing   162
    Cleaning of Systems   162
    Fire Protection Systems   164
    References   165

Chapter 7. Water Treatment   167

    Zeolite Water Softening   173
    POU Methods of Treatment   175
    Nitrate Contamination   175
    Water Treatment Methods for Producing High-Purity Water   175
    Iron and Manganese Removal   180
    Gadgets   181
    Chemical Feed Systems   182
    References   183

Chapter 8. Materials   185

    Quality of Work   186
    Stainless Steels   187
    Copper-Bearing Alloys   188
    Protective Coatings   188
    Plastic Piping   189
    Corrosion of Valves   192
    Corrosion of Pumps   193
    Pump Mechanical Seals   194
    References   194

Chapter 9. Building Systems and Maintenance   197

    Domestic Hot Water   197
    Materials in Water-Using Systems   198
    Monitoring   199
    Survey of High-Rise Buildings   200
    References   203

Appendix A. Water Treatment Specifications 205

Appendix B. Specifications for Water-Testing Reagents and Equipment 225

Appendix C. Sampling and Methods of Water Analysis 237
Sampling 237
Water Testing and Methods of Water Analysis 239
Instrumental Test Methods 256

Appendix D. Computer Programs 263
CORSCNEF 263
WTRCHEM84 268

Index 275

Preface

The purpose of this book is to encourage architects, building designers, and physical plant operators to recognize the importance of carefully designing and properly maintaining building water systems. The reader of this book will gain a full understanding of the principles of water treatment and the role of proper material selection and as a result should be able to prevent the development of corrosion and scale problems. The book should prove helpful to both chemists and operators and particularly to those just starting to get involved with water chemistry and water treatment.

I was fortunate to have been chosen to initiate an institutional feedwater treatment program under the direction of Drs. A. M. Buswell and T. E. Larson of the Illinois State Water Survey. Specifications were prepared for purchasing water treatment and testing equipment on a competitive bidding basis, and treatment guidance was provided for more than one hundred Illinois institutions. Cooperation by all state agencies was excellent, and this program has continued to function for more than fifty years in an efficient manner. This has enabled field-type research to be conducted, resulting in the presentation of papers and the development of patents. Much of the information presented in this book has been derived from this treatment program.

This book is considered to be practical rather than theoretical. Accent has been placed on collecting much of the information and references needed for solving corrosion and scale problems. In the early days of my water-treating experience (before the Langelier Saturation Index), pseudoscience may have been the vogue, but during my lifetime, water treatment has developed into a mature science. This book presents the up-to-date, modern techniques, which are now available to the present-day water chemist, for solving the corrosion and scale problems in industries, in buildings, and in the home.

I have found water treatment to be an interesting field for more than fifty years, and I intend to keep active in it as long as I am physically able. Solving corrosion and scale problems has been stimulating, and the sociability with other chemists and operators has made it particularly rewarding. Sharing water treatment information and experiences with other water treaters at water conferences has been most satisfying. It has also been most gratifying to have operators develop a real interest in water treatment and to solve their water treatment problems conscientiously through friendly and cooperative effort.

In a way, this book serves as a story of my life, as water treatment has been one of my chief concerns, along with my family and travel, made possible by the rewards from water treating. Thanks are owed to my wife and daughters for putting up with the countless hours I have spent in travel, in the laboratory, in front of the computer, and in writing this book.

Russell W. Lane

Chapter

1

Introduction

Water in building water systems is exposed to many metals and to different temperature, velocity, and pressure conditions. This book will discuss the effects of these variables on the metals in these systems and will provide information for correcting water-caused problems in building systems.[1] It will provide architects, building designers, managers, and operators with a better understanding of:

1. The importance of water quality and treatment in preventing and controlling corrosion and scale
2. The possible need for installing water treatment equipment for application of water treatment chemicals and for monitoring results
3. The importance of selecting the proper materials of construction

In modern building water systems, there are dozens of different uses of water, such as chilled water for drinking; domestic hot water for lavatory and laundering; water for fire protection; water for aquariums, pets, and plants; and water for the physical plant processes of humidification, cooling, and heating. These separate systems, such as the cooling and heating systems, may be particularly complex, often involving many open and closed systems. Each of these systems must be considered specifically in planning the selection of materials and the required water treatment (if necessary) in order to avoid subsequent corrosion and scale problems. The differences in temperature, pressure, and velocity encountered in the different uses of water in a building affect its properties with respect to scaling and corrosive tendencies.

Therefore, the technology learned from theory and experience must be applied in determining the correct selection of materials and method of water treatment. As would be expected, boiling water under pressure in a boiler requires a different metal of construction and a different method of water treatment than in the case of chilled potable water.

For example, domestic hot water used in the lavatory and laundry may require a different metal and water treatment method; however, this isn't as complex as it may first sound, since usually one only needs to know the metals to avoid rather than to install. Economics may likely dictate that there be only two or three metals from which to choose. In open and closed cooling and heating systems, economics will usually dictate that these metals be steel and copper-bearing metals (for cooling system heat exchangers). Since steel is subject to corrosion by water, accent must be placed on designing a proper water treatment so that corrosion control is adequate to ensure that piping and water-handling equipment are maintenance-free and of long life.

The design of water treatment for scale control is based practically entirely on the analysis of the water supply, and the decision as to what metal to install with respect to scale control does not need attention.

Before new buildings are designed and constructed, it would prove beneficial to all concerned if the architect and/or building designer would contact water treatment or corrosion engineers to obtain their advice on the proper materials of construction and treatment (if necessary) for the particular water supply to be used. These experts have extensive knowledge and experience and have observed which materials best provide good service and last longest when exposed to waters of different qualities under the expected variable environmental conditions.

What is involved in maintaining proper corrosion and scale control in a water system? Corrosion, best described as the degradation of a metal by an electrochemical reaction, reveals its importance to the nation in its annual environmental costs, estimated at $142 billion (1988 dollars).[1-3] Corrosion in water systems represents a good portion of those costs. Although water supplies are reported to be noncorrosive, *all water is corrosive*. Differences noted are purely in degree.

Some supplies require more corrosion-resistant metals and more costly water treatment than others. Many supplies require conventional metals of construction and little or no additional water treatment. Even the *same* water can require a different choice of metal or water treatment for a particular use. Many water supplies used at room temperature conditions and exposed to conventional metals of construction require little or no additional water treatment besides that applied by the water purveyor.

There are many types of corrosion resulting from electrochemical reactions. The various types of corrosion are discussed in detail in Chap. 2, which is on the subject of corrosion and methods for inhibiting and controlling corrosion.

How does the water user see the effects of corrosion? The water may appear discolored or dirty; it may have an off-taste or odor; it may plug pipelines with scale or corrosion products or cause leaks to develop from perforation of piping due to corrosive attack; or it may lower heating or cooling efficiency.

There are *simple tests* for determining whether corrosion or scaling is occurring and for identifying the scale or corrosion products. The more common corrosion products observed are brown iron oxide (Fe_2O_3) (hematite) or black iron oxide (Fe_3O_4) (magnetite) in the case of steel corrosion and blue or blue-green to black copper oxides in the case of copper-bearing metal corrosion. Corrosion products may be identified by dissolving a small quantity of the deposit in hydrochloric acid (1:1) and performing a qualitative test for iron, copper, or otherwise (see App. C). The black iron oxide, or magnetite, may be identified specifically by its magnetic characteristics. Scale (usually calcium carbonate) can be identified by noting effervescence (carbon dioxide release) when dilute hydrochloric acid (1:1) is added to the sample. Other scales, such as calcium sulfate and calcium phosphate, may be identified by testing the acid solution with appropriate reagents. A complete analysis of the deposit obtained by submission of a sample to an analytical laboratory may be required to obtain a more exact diagnosis for correcting the corrosion or scale problem.

All corrosion problems could have been avoided by the proper initial selection of materials and/or water treatment. Some water supplies may be more difficult to treat properly to correct these problems than others, but *all* such problems can be corrected with adequate effort and expenditure.

Off-flavors may result from corrosion of metals within the systems whereby the dissolved metal content exceeds the *MCL (maximum contaminant level) specified in the U.S. Drinking Water Standards,* shown in Table 3.1. In recent years, considerable concern has been directed toward the purity and contamination of drinking water, particularly with respect to lead as a contaminant derived from corrosion of metals in building piping systems. Another concern has been the organic contaminants in municipal supplies derived from surface water supplies and their treatment. These matters are discussed in detail in Chap. 3.

When we speak of scale, or water-formed deposits or sludge, usually we mean a mineral deposit that coats heat transfer surfaces when water is heated. The most common deposit is calcium carbonate, which precipitates from solution under certain conditions of water quality (hardness, alkalinity, and pH) when water is heated. The experienced water chemist, being aware of the composition of the water, can predict the need for water treatment for control of corrosion and scale and can prescribe the necessary treatment and equipment. By calculation of various water indexes, the water chemist can estimate the corrosiveness or scaling tendency of a water. Methods of calculating these indexes is also covered in Chap. 3.

It is necessary to provide a means for reducing the scaling tendency of potentially scale-forming waters, particularly in domestic hot water systems; otherwise, deposits causing reduced flow and lower hot water temperatures, as well as turbid and discolored water, may result. The most common deposit, hardness (calcium carbonate) scale, usually appears to the observer as an off-white (perhaps rust-tinted) hard and crystalline material on the piping surface. A less common deposit—a white, slimy, insulating-type (magnesium hydroxide) scale—may result when water is lime-softened by the water purveyor and the softened water has not been properly treated or adjusted to a lower pH before filtration and discharge to the city mains. Resulting scale is shown in Fig. 1.1.

Heating domestic hot water in "instantaneous heaters" or multiple-tube heat exchangers or resident home furnaces in which water velocity may be lowered may cause sediment formed from water hardness to settle out, with subsequent lowered flow, temperature, and water quality. Scale may cause the overheating and burnout of electric-resistance type water heater elements or failure of thermostatically controlled hot water mixing or tempering valves. These are some of the bad effects produced by scale-forming waters. Scale inhibition is discussed in detail in Chaps. 3 and 7.

Figure 1.1 Heavily scaled piping. (*Illinois State Water Survey, Champaign, Ill.*)

Water and Its Properties

Since we are to be concerned with water and its properties throughout this book, this is a good time to discuss water and its basic chemistry in order to better understand water's behavior. Water, a most essential part of our life on earth, has a simplified structure consisting of two atoms of hydrogen and one of oxygen in the molecule (H_2O). Actually, its structure is much more complex than this, which may explain its many unusual physical properties, such as high surface tension and dielectric constant and its property of expansion on freezing.[3-5] Because of this latter property, ice floats on open bodies of water; water does not freeze from the bottom up, which would prevent life, as we know it, from existing.

Water is not found in its pure form on earth, as even rainwater dissolves gases and dusts in the atmosphere. It covers three-fourths of the earth's surface and as the universal solvent, it dissolves almost everything it contacts. As a result, water supplies vary in content throughout the world.

Water ionizes only very slightly, yielding only 10^{-7} mol of hydrogen and 10^{-7} mol of hydroxyl ions per liter; therefore, it cannot conduct electricity in its pure form. As salts or ionizing materials dissolve in it, the electrical conductivity increases and can be used as a measure of its dissolved mineral content.[4,6] The ion product of water ionization, a measure of the degree of formation into H^+ and OH^- ions, is the square root of K_w, which is equal to 10^{-7} at 25°C (77°F) and indicates that only 1 out of 555 molecules is ionized. The pH of an aqueous solution is equal to the log of $1/H^+$, where H^+ is the concentration (actually activity) of the hydrogen ion. An aqueous solution is considered neutral when H^+ equals OH^-.

The Composition of Waters

The analytical content of waters is important to the user and the water treatment consultant in determining the methods of treatment to be applied to correct properties that may cause health, corrosion, and scale problems, as well as interruptions and inefficiencies in manufacturing processes.

For example, knowledge of the degree of hardness of a water supply may disclose that serious scale may be produced in its use in industrial processes. The hardness of water is expressed as calcium carbonate (as $CaCO_3$) as a common term rather than specifically as calcium bicarbonate, magnesium bicarbonate, calcium carbonate, calcium sulfate, magnesium chloride, or otherwise. For example, 100 mg/L of calcium sulfate hardness is converted to hardness (as $CaCO_3$) as follows:[6]

$$100 \times (\text{mol wt of CaSO}_4/\text{mol wt of CaCO}_3)$$
$$= 100 \times 136/100$$
$$= 136 \text{ mg/L (as CaCO}_3)$$

The composition of waters in the United States varies greatly, and examples of this variability are noted in the following information. The average hardness content (as calcium carbonate) of the water supplies of major cities varies from 7 to 285 mg/L, as shown in Table 1.1.

In general, deep well-water supplies are higher in hardness, alkalinity, and electrical conductivity (a measure of the dissolved solids content) than waters derived from rivers and mountain runoff, as may be found on the east and west coasts. In the midwest, waters are generally higher in hardness, alkalinity, and conductivity, particularly those derived from wells. Surface water supplies from rivers and lakes are usually not as high in hardness, alkalinity, and conductivity but may be less palatable. This variation in analyses should make one realize why the chemical composition of different water supplies is so important and has such a genuine effect on the advice for proper treatment provided by a water treatment expert. While the hardness content of a water supply is informative, the water chemist must consider other constituents—such as calcium, chloride, sulfate, iron, manganese, silica, dissolved oxygen, pH, total dissolved and suspended solids, and organic contents—in evaluating the corrosive and scale tendencies of a water supply and in deciding on the necessary water treatment and the proper materials of construction for the water system.[6,7]

The water consultant, in recommending proper water treatment to correct a scale or corrosion problem, will emphasize the importance of accurate proportionate application of the necessary chemicals. In addition, conducting accurate periodic water tests according to established test procedures[7-9] will be recommended. Such testing needs to be done

TABLE 1.1 Average Hardness Content of Water Supplies of Major U.S. Cities, mg/L

City	Hardness (as CaCO$_3$)	City	Hardness (as CaCO$_3$)
Boston	13	Indianapolis	285
New York City	36	Chicago	140
Miami	61	Denver	78
New Orleans	91	Los Angeles	110
Dallas	90	Seattle	20

SOURCE: S. Smith, "A Geographical Look at Water Quality in the US," *Water Technology,* November 1988, p. 33.

consistently and regularly and the results used in adjusting treatment to attain the continuous recommended treatment levels. Water test reports need to be submitted to a knowledgeable supervisor or consultant regularly for review and recommendations. In addition to the preparation of water test reports, corrosion test coupons should be installed to determine whether the treatment recommendations are accomplishing the desired results. The consultant should also make periodic plant visits and analyze samples taken periodically to be assured that treatment plans are being effectively implemented.

Expression of Concentrations of Water Constituents

Concentrations of water constituents are expressed as follows:[6]

ppm = mass of substance/mass of solution
(for example, 1 pound per million pounds, or 1 mg/1 kg)

mg/L = mass of substance/volume of solution (liter)
(for example, 1 mg/1 L)

As we may be dealing mainly with water of low solids content, its density or specific gravity is near 1.0; therefore, we can consider 1 ppm = 1 mg/L. If sea water (>30,000 ppm total solids or brackish waters of about 10,000 ppm) is analyzed, then the specific gravity must be taken into consideration in converting milligrams per liter to parts per million, as parts per million will not equal milligrams per liter and corrective calculations are necessary. Since parts per billion (ppb) (equivalent to 1/1000 ppm) are also of concern, micrograms per liter (μg/L) can be generally considered to be equal to parts per billion (ppb). Milliequivalent per liter (mequiv/L) is also a common term used to express concentrations. Since the equivalent weight (equiv wt) of Ca^{2+} is 40 g/mol and there are 2 equivalences per mole for calcium, then 20 g is the equivalent weight; with 100 mg/L Ca^{2+}, 100 divided by the equivalent weight of 20 = 5 mequiv/L.

The old term *grains per gallon* (1 gpg = 17.12 ppm) is still used mainly with respect to water hardness and to ion exchange capacities [for example, 20,000 gr (as $CaCO_3$)/ft^3].

In building water systems, the hardness, alkalinity, pH, and conductivity are usually the most meaningful constituents of a water supply with respect to its corrosive and scale-forming properties. Of course, the potability of a supply must be assured, but it is assumed in most cases that the water supply is derived from a municipal supply that has been properly treated to render it safe for drinking. In later chapters, attention is directed toward the installation of suitable backflow prevention devices to avoid any possibility of nonpotable supplies becoming mixed with the potable supply.

Most Common Causes of Corrosion and Scale Problems

The most common causes of corrosion and scale problems in interior water systems are the following:[1,6,7]

1. The complete softening of the entire water supply. This generally causes an increased corrosiveness of the supply. However, certain equipment and places, such as boilers, laundries, laboratories, and open and closed systems, may require completely softened water in order to keep scale formation under control. This is discussed in detail in later chapters.
2. Choice of the wrong piping material or a combination of the wrong materials, resulting in piping failures caused by corrosion.
3. Failure to consider the effect of velocity on the choice of piping material and in sizing of the piping.
4. Failure to install appropriate water treatment equipment and to provide easy access for repair or replacement (or to apply proper water treatment chemicals).
5. Operation of domestic hot water at too high a temperature (>135°F).[7,10]
6. Failure to apply recommended chemicals for treatment of domestic hot water, boilers, cooling towers, and closed systems and to exert daily conscientious attention to treatment control.
7. Failure to arrange for a corrosion testing location where corrosion coupons can be installed for observing and monitoring the piping system to determine whether chemical treatment or changes in treatment are needed.
8. Failure to design a new system so that it can be adequately cleaned, passivated, and flushed before general usage.
9. Failure to consider the effect of original building design on the design of subsequent additions.
10. Failure to adequately inspect original plumbing installations to be assured that:
    a. Fittings that reduce flow are not installed.
    b. Noncorrosive flux or pipe compounds are used judicially.
    c. Reaming of pipe and soldering were done competently, resulting in piping in which water will flow with minimal turbulence.

11. Allowing stagnant water to exist in a new building for several months before tenant occupation. In Australia and New Zealand, the serious problems with pitting of copper tubing in new buildings were determined to be caused by this practice.

References

1. *Prevention & Control of Water-Caused Problems in Building Potable Water Systems,* TPC Publication 7, National Association of Corrosion Engineers, Houston, 1980.
2. *Economics of Internal Corrosion Control,* American Water Works Association Research Foundation, Denver, 1989.
3. *Barnstead Basic Book on Water,* Barnstead Co., Boston, 1971.
4. F. Kemmer, *Water: The Universal Solvent,* 2d ed., Nalco Chemical Co., Oak Brook, Ill., 1979.
5. C. A. Hampel and G. G. Hawley, *The Encyclopedia of Chemistry,* 3d ed., Van Nostrand Reinhold, New York, 1973.
6. C. E. Hamilton (ed.), *Manual on Water,* ASTM Special Technical Publication 442A, American Society for Testing and Materials, Philadelphia, 1978.
7. R. W. Lane, *Industrial Water Treatment Guidelines and Water Analytical Methods,* Champaign, Ill., 1983.
8. *1991 Annual Book of ASTM Standards,* vol. 11.01: *Water,* American Society for Testing and Materials, Philadelphia.
9. American Water Works Association, *Standard Methods for the Examination of Water and Wastewater,* 17th ed., American Public Health Association, Washington, D.C., 1989.
10. A.S.S.E. *Energy & Water Conservation Guidelines for Plumbing Systems,* American Society of Sanitary Engineers, Bay Village, Ohio, 1987, chap. 3.

Chapter 2

Corrosion

Corrosion touches on all of our lives in the form of the rusting of the steel in our cars, the metal failures of appliances and external home surfaces, and many other occurrences, such as failures of bridge and highway structures. There have been a number of cases in which lives have been lost owing to metal failures caused by corrosion.

In water systems, the annual cost of corrosion of distribution piping and of home plumbing and fixtures in the United States has been estimated to exceed $700 million.[1] In building water systems, corrosion of metals is of more concern than in the past, since component metals such as lead and cadmium, which are highly toxic, have been found to be dissolved sufficiently in corrosive drinking water to provide serious health hazards. The reported effects of lead include adverse effects on the brain and nervous system, the reproductive system, the circulatory system, and the kidneys. The corrosion of copper, iron, and zinc may cause perforation of piping, and corrosion products of these metals may cause plugging of water lines, colored stains on laundered items, stains on lavatory equipment, and off-flavored water. Chapter 3 elaborates on these matters and includes information on metallic contamination.

Common evidence of corrosion is the rapid formation of iron rust (a red-brown or black discoloration); it will form on an unprotected steel surface in contact with water or humid air. Water in contact with the rusted surface may also show discoloration. These hydrous iron oxides are usually in the form of red-brown ferric oxide (Fe_2O_3) (hematite) or black ferric oxide (Fe_3O_4) (magnetite), the latter of which may be identified by its distinctive property of being magnetic.

Corrosion—An Electrochemical Process

Since the corrosion process has been identified as an electrochemical reaction, anodes (positive poles) and cathodes (negative poles) are involved. The most common illustration or example would be two dissimilar metals such as steel and copper being connected as steel piping and a copper-bearing metal valve. In this case, the steel would be the anode and the copper the cathode, and as a result, the steel pipe threads would be observed to corrode, and copper would be protected. In the case of galvanized steel, zinc serves as the anode, or the sacrificial metal to inhibit the corrosion of the cathode, steel. Another description of the anode would be its serving as the electrode where loss of electrons (flow of electricity) occurs, positive metal ions are formed, and corrosion and chemical oxidation occur.[2–4]

It should be recognized that the corrosion cell is made up of four components. Two that we have already discussed are the anode and the cathode; however, the electrolyte—water, for example—and the electric circuit are also essential for corrosion to progress. In order to reduce or stop corrosion, the designer may make use of these components by:

1. Insulating the anode from the cathode
2. Applying a protective coating to interrupt the corrosion current
3. Imposing a countercurrent electric current against the corrosion current (essentially cathodic protection)
4. Eliminating the electrolyte (keeping the metal dry)

On a corroded steel surface, numerous small individual anodes and cathodes may form to perpetuate the corrosion process. Such couples are formed by slight differences of exposure in the environment, as when they are imperfect protective surface films, and by slight differences in air and water contact and in cleanliness of the surfaces.

The corrosiveness of a water depends entirely on its degree of saturation with the ions or molecules of the metal or compound with which it is in contact. The rate of corrosion depends on the reactions at the anode and cathode and can be measured by a corrosion current. This rate (or corrosion current) is dependent on the electromotive force involved (see Table 2.1), the resistance of the electrolyte, and the particular metals and metallic films involved. Since there is little information available on the resistance of different metallic films and rates of reaction, corrosion rates are usually most reliably determined by exposing the specific metal (coupon) to the particular environment in question for set time periods.

Corrosion testing can be conducted by determining the weight loss over an exposure period or by determining the corrosion current as specified in ASTM D2776.[5] Since the corrosion rate in building drinking water systems is usually rather slow and may involve the development of a bulky precipitate as the protective or inhibitive mechanism,

TABLE 2.1 Galvanic Series of Metals and Alloys in Flowing Aerated Seawater at 40–80°F (4.4–26.7°C)

Corroded End (Anodic, or Least Noble)
Magnesium alloys
Zinc
Beryllium
Aluminum alloys
Cadmium
Mild steel, wrought iron
Cast iron, flake or ductile
Low-alloy, high-strength steel
Ni-resist, Types 1 & 2
Naval brass (CA464), yellow brass (CA268), aluminum brass (CA687), red brass (CA230), Admiralty brass (CA443), manganese bronze
Tin
Copper (CA102, 110), silicon bronze (CA655)
Lead-tin solder
Tin bronze (G & M)
Stainless steel, 12–14% Cr (AISI Types 410, 416)
Nickel silver (CA732, 735, 745, 752, 764, 770, 794)
90/10 Copper-nickel (CA706)
80/20 Copper-nickel (CA710)
Stainless steel, 16–18% Cr (AISI Type 430)
Lead
70/30 Copper-nickel (CA715)
Nickel aluminum bronze
INCONEL* alloy 600
Silver braze alloys
Nickel 200
Silver
Stainless steel, 18% Cr, 8% Ni (AISI Types 302, 304, 321, 347)
MONEL* alloys 400, K-500
Stainless steels, 18% Cr, 12% Ni-Mo, (AISI Types 316, 317)
Carpenter 20† stainless steel, INCOLOY* alloy 825
Titanium, HASTELLOY‡ alloys C & C276, INCONEL* alloy 625
Graphite, graphitized cast iron
Protected End (Cathodic, or Most Noble)

*International Nickel trademark.
†The Carpenter Steel Co. trademark.
‡Union Carbide Corp. trademark.
SOURCE: *HVAC Applications Handbook,* American Society of Heating, Refrigeration and Air-Conditioning Engineers, Atlanta, 1991, p. 43.2.

weight loss procedures as specified in ASTM D2688 usually prove most informative. It is recommended that the method of monitoring described in ASTM D2688 Test Method B be installed in buildings in which the water quality is expected to be questionably corrosive or scale-forming. By conducting periodic tests, the building owner or manager can become aware of possible scale or corrosive conditions developing and can arrange for changes in water treatment equipment or chemicals before serious malfunctions occur in the form of leaks and piping plugging with corrosion products.[5]

Since metals are commonly derived from natural mineral ores, they have an inherent tendency to revert to the stable form in which they were found originally in the earth. Different metals and the components of alloys have greater or lesser tendency to revert to their natural forms. Their relative electrochemical potential (to the hydrogen electrode) or oxidizing or reducing characteristics at the water interface determine their tendency to corrode. This is particularly of concern when two dissimilar metals (such as iron and copper) are in contact with the same environment.

Table 2.1 shows that the top (less noble) metals have a much greater tendency to corrode than the lower (more noble) metals. Accordingly, it would be expected that zinc in combination with steel would corrode more readily; in fact, it would provide cathodic protection for steel, as evidenced in galvanized steel. The farther apart the metals stand in the series, the greater the galvanic tendency. For effective cathodic protection, a metal such as zinc (the anodic metal) should have a large relative surface area in contact with steel (the cathodic metal) rather than a small area, since the steel is protected at the expense of the zinc where steel is exposed at joints and holidays. This explains the expected longer life of galvanized steel compared with steel.

The corrosion of iron may be illustrated from the reaction of iron with water as follows:

$$Fe \rightarrow Fe^{2+} + 2e^{2-}$$

where the iron goes into solution and 2 electrons flow through the metal from the anode to the cathode and the following reactions occur:

$$2H_2O + 2e^- \rightarrow 2OH^- + H_2 \text{ (gas)}$$

$$3Fe + 4H_2O \rightarrow Fe_3O_4 + 4H_2O$$

In the absence of dissolved oxygen, hydrogen production is the rate controlling reaction.

In the presence of dissolved oxygen, the thin film of magnetite is oxidized to hydrated iron oxide, which is insoluble in water and precipitates on the metal surface, possibly forming tubercles of iron oxide and restricting the flow of oxygen to the surface, as follows (see Fig. 2.1):

$$\tfrac{1}{2}O_2 + Fe_3O_4 \rightarrow 3Fe_2O_3$$

An explanation for the aggressiveness of chloride and sulfate ions is their ability to solubilize ferrous iron and to form H^+ ions (acid) in re-

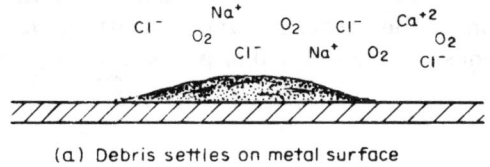

(a) Debris settles on metal surface

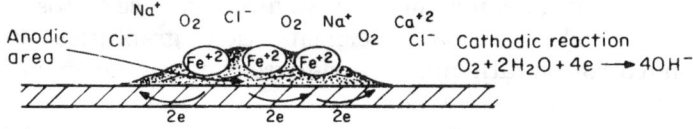

(b) Oxygen can reach metal only at open surface.

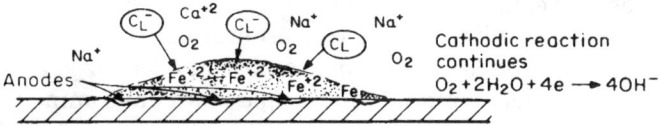

(c) Oxygen continues to depolarize the cathodic area while chloride diffuses into the porous deposit.

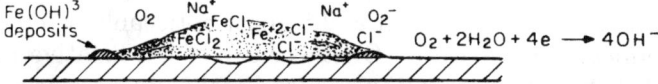

(d) The iron within the deposit remains soluble as Fe^{+2} in the absence of O_2; and corrosion increases as ionic strength in the deposit increases.

Figure 2.1. Successive steps in the formation of an oxygen concentration cell as a consequence of deposit on a steel surface in oxygenated water. [*Frank N. Kemmer (ed.), The NALCO Water Handbook, 2d ed., McGraw-Hill, New York, 1988.*]

action with water at the anode, thus causing a potential gradient between the anode and cathode and resulting in a corrosion cell.

The surface of metal is often covered with small anodes and cathodes if the protective oxide film is thin and continuous, providing *general, or uniform, corrosion*. This is more desirable than a metal surface that is only partially coated with a protective oxide film, particularly if the anode area is small and the cathode large. Under these conditions, serious pitting and deep penetration may occur in the small anode area. A large anode area and smaller cathode area provide less serious corrosion.

The rate of corrosion is determined by the protectiveness of the corrosion products formed, the degree and kind of corrosive elements present (such as dissolved oxygen, pH, chloride, and sulfate), the temperature, the configuration of the structure and velocity of the environment, and the effectiveness of the natural or added corrosion inhibitor.

Types of Corrosion[6]

There are a number of different types of corrosion, namely, uniform (general), pitting, galvanic, concentration-cell [crevice corrosion or MIC (microbe-induced corrosion)], dezincification, graphitization, stress corrosion (corrosion fatigue), erosion-corrosion (impingement and cavitation types), and stray current.

1. *Uniform, or general, corrosion* is recognized as taking place at a generally equal rate over the entire surface and may be best described as the corrosion resulting from acids in a water environment having minimum and not localized protective properties.

2. *Pitting corrosion* is nonuniform, occurs at a localized anodic area, may be sharp and deep, and is an example of an environment offering some protective properties but not complete corrosion inhibition. It is associated with concentration-cell corrosion, galvanic corrosion, and crevice corrosion.

3. *Galvanic corrosion* is the result of the exposure of two dissimilar metals in the same environment and is most noticeable when they are directly connected electrically. On the basis of the relative potential (Table 2.1) of the two metals, the one less noble will corrode at the expense of the other more noble metal, thus offering protection for the cathodic metal. An example is the corrosion of steel piping near the more noble metal copper found in valves. Galvanic corrosion is increased by a greater difference in potential, by increased closeness of the metals, and by increased mineralization or conductivity of a water.

4. *Concentration-cell corrosion,* which is probably the most prevalent type of corrosion, occurs when there are differences in mineralization, acidity, metal-ion concentration, anion concentration, dissolved oxygen, and temperature in exposure of a metal to its environment. These differences cause differences in the solution potential of the same metal.

5. *Crevice corrosion* is an example of concentration-cell corrosion in which oxygen becomes deficient in the crack or crevice; this causes a difference in potential and results in corrosion. The most obvious example is the case of dirt or debris precipitating on a metal surface and causing a difference in oxygen diffusion to the metal surface. This causes a difference in potential to develop under the dirt and between the surface under the dirt and the nearby clean surface (differential aeration) and results in corrosion occurring under the deposit. Under-

deposit corrosion is another example in which deposits—bacterial growth, dirt (from dust in the air), and suspended matter—adhere to a metal surface, forming an electrolytic cell between the area under the deposit and the clear area next to it. The area under the deposit becomes anodic to the clear area and is corroded; often a pit is formed under the deposit. Keeping the bottom of a cooling tower basin (a particularly vulnerable area where crevice corrosion can develop) free of sludge is important in diminishing these corrosive tendencies. The installation of bypass filters to filter part of the total cooling water flow continuously often proves worthwhile in reducing these deposits and subsequent corrosion. Well-engineered filter systems that automatically periodically backwash the filters and provide filtration of the total flow are available from various manufacturers and would be particularly recommended for installation in areas where dust is a problem. These systems may have to be bypassed during periods when neighboring trees shed their blossoms and cause rapid plugging of the filters.

Providing clean surfaces of metal is of benefit in reducing corrosion, particularly if the metal has just been installed and is partially coated with a protective oil-type coating. Rinsing first with an alkaline surface-active solution to remove oil and then using a passivating agent, such as sodium acid phosphate, provides a cleaner surface. This will result in freedom from partially protective film surfaces, which encourage the development of small individual anodes and cathodes. A clean surface is particularly required for proper maintenance of stainless steel because it is necessary for it to have ready access to oxygen in the air for preservation of its corrosion-free property.

6. *Dezincification corrosion* occurring in a copper-zinc alloy, such as brass, is the result of zinc being more anodic than copper and being corroded in a hostile environment, leaving the copper in situ. Yellow brass in soft, unstable waters is particularly subject to this type of corrosion; however, red brass and Admiralty metal containing less zinc are much less subject to this type of corrosion.

7. *Graphitization corrosion* is a form of corrosion of cast iron that occurs in mineralized waters or waters of low pH. Graphite dispersed in cast iron serves as the cathode, and the iron-silicon alloy, the anode. This results in dissolution of the iron alloy and leaves black, spongelike graphite as a structurally deficient material.

8. *Stress corrosion* (*corrosion fatigue*) results from external tensile stress and is usually in evidence at the microstructure grain boundaries of the metal. Repeated rupture of the protective film on the surface often provides a continuously anodic region, and the result is cracking and failure of the metal. Stress corrosion cracking (SCC) is evidenced in caustic embrittlement of steel boiler tubes and drums and in chloride attack of stainless steels, for example. In the case of boiler wa-

ters, high levels of causticity and lack of necessary corrosion inhibitor concentrations accompanied by stress can cause intergranular or transgranular attack of steel and can result in metal rupture. The stress corrosion cracking of austenitic stainless steels (such as 304) exposed to chlorides is a common example of the susceptibility of stainless steel to corrosion.

9. *Erosion-corrosion (cavitation corrosion)* is the result of the continuous removal of the protective film of corrosion products, which serves as the barrier to corrosive attack of some metals. Impingement attack of copper tubing is a good example; exposure of copper to velocities greater than 1.2 m/s (4 ft/s) and sudden changes in direction of flow causes serious corrosion in the form of deep, rounded holes, as shown in Fig. 2.2. With cavitation corrosion, high velocity and changes in direction of flow result in gas-bubble formation at low pressure points and resolution of the gases at high pressure points. Wiredrawing of faucet seats, in which grooves appear across the face of the seat, is another example of cavitation or erosion-corrosion.[7]

10. *Stray current corrosion* may be blamed for corrosion occurrences when other causes are responsible. An example where stray currents may be the cause of corrosion would be underground sites near direct

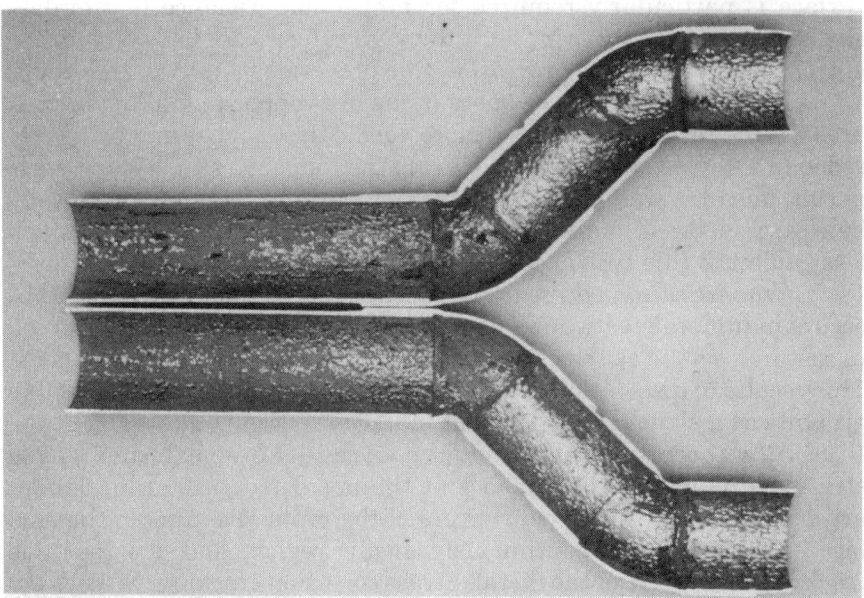

Figure 2.2 Erosion-corrosion of copper. (*Illinois State Water Survey, Champaign, Ill.*)

current sources from subway trains. A present common example of stray current corrosion is observed in piping that is near piping systems protected by cathodic protection systems. The stray currents derived from these cathodic protection systems may cause severe corrosion of these nearby systems when current leaves the close-by systems. The installation of insulating couplings or proper countercurrent applications such as cathodic protection may provide a means for counteracting this current flow problem.

Cathodic protection is used effectively in protecting underground piping from corrosion by soils. In this case, buried anodes located at proper intervals provide the necessary current to counteract corrosion currents. Transcontinental oil and gas pipelines make effective use of cathodic protection to reduce potential leaks and to extend the life of underground piping. Another effective use is to mount plastic-backed sacrificial anodes next to the steel tube sheets of heat exchangers so that the steel next to the copper-bearing metal tubes will be protected from corrosion.

In the author's opinion, the installation of dielectric unions used for insulating dissimilar metal fittings from each other is best justified when the electric conductivity of the water exceeds 500 µS/cm. Waters that have high dissolved solids contents and contain high chloride and/or sulfate can cause increased galvanic corrosion between galvanized piping threads and brass valves to occur within 5 years, while low-conductivity waters will not likely cause such failures before 20 years.

Attention is directed particularly toward the care of equipment during outages, such as in the summer and winter. Metals exposed to *stagnant* untreated water are subject to increased corrosive tendencies and must therefore be given special attention at these times. The lay-up procedures for air-conditioning and boiler systems described in Chaps. 4 and 5 may serve as general guidelines for proper lay-up of equipment.

Anodic, Cathodic, and Mixed Inhibitors

In speaking generally of corrosion inhibitors, we have anodic inhibitors, which are effective at anodes; cathodic inhibitors, which are effective at cathodes; and mixed inhibitors, which include both anodic and cathodic inhibitors and may be effective at both electrodes. Examples are chromate and nitrite as anodic inhibitors, including the precipitating film type as orthophosphate and silicate. Cathodic inhibitors are, for example, zinc, polyphosphate, and carbonate alkalinity. Mixed inhibitors, as stated above, are a combination of both anodic and cathodic inhibitors. Organic inhibitors, such as soluble and dispersible oils, are also used. The sodium salts of mercaptobenzothiazol, benzothiazol, and tolyltriazole are included in most cooling water formulations to provide corrosion inhibition of copper-bearing metals.

Water Constituents and Properties— Important in the Corrosion Process[8]

Water constituents that may influence the corrosion rate of metals in water distribution systems are calcium, magnesium, sodium, chloride, sulfate, dissolved oxygen, alkalinity, pH, buffer intensity, free chlorine, chloramine, suspended solids, and conductivity (dissolved solids).

In general, waters of the following properties will tend to be more corrosive:

1. Softer waters (below 60 mg/L hardness).
2. Waters lower in pH and alkalinity. The addition of calcium, alkalinity, and chemicals for pH adjustment (such as lime [$Ca(OH)_2$], carbon dioxide, and/or soda ash) to attain a desired pH and alkalinity in treated waters has been found effective in correcting the corrosiveness of low hardness–low alkalinity waters.[8]
3. Waters high in chloride and/or sulfate (>150 mg/L).
4. Waters containing appreciable dissolved oxygen.
5. Waters low in buffer intensity (also sometimes called buffer capacity). Such waters tend to be more corrosive owing to their lack of buffer intensity to counteract local acid production in the corrosion cell. Buffer intensity is defined as the moles per liter of strong base (or OH^-) which, when added to a solution, causes a unit change in pH. A water of at least 0.5 mequiv/L buffer intensity will have sufficient buffer capacity to neutralize local acid concentrations generated at the cathode, and as a result, the corrosion cell reaction will be stifled rather than promoted.[9,10]
6. Waters of low pH (<6.0) and high conductivity (>500 µS/cm).
7. The presence of free chlorine above 1 mg/L and chloramine above about 2 mg/L.
8. The presence of suspended solids (such as sand or crud). These solids tend to increase the erosive or erosion-corrosion tendencies of a water and also provide dirt (crud) for forming deposits in the water lines. A deposit of crud may lead to concentration-cell corrosion, which results from the difference in the oxygen concentrations underneath and next to the deposit.

Corrosion Testing[5]

Test Method B described in ASTM D2688 provides the most informative results in testing the corrosiveness of steel, galvanized steel, and copper in building water systems. Machined pipe specimens 4 in long prepared from the same pipe as in the piping system under test are reduced in metal thickness to provide an acceptable weight for weighing on laboratory balances. After the exterior surface is painted in order to eliminate all corrosive areas except the internal surface, the specimens are then installed in plastic sleeves to provide insulation from other metals in the water system. Design is arranged so that streamline flow is provided without turbulence throughout the test. Specimens are weighed before painting and installation and again after 6 to 24 months of exposure. Weight losses are then determined after the oxide layer is removed and the corrosion rates calculated. These rather lengthy periods of time for exposure are necessary, since it is recognized that the effective protective oxide film is slow to build up; so to obtain an accurate evaluation of the corrosion rate, sufficient time must be allowed. Such protective oxide films are typical of the corrosion occurring in building water systems. Figures 2.3, 2.4, and 2.5 show the test assembly and parts thereof.

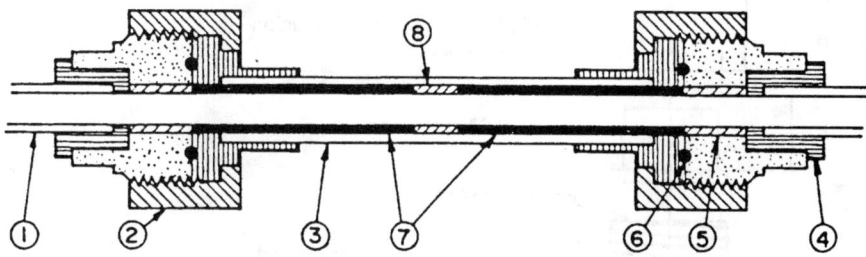

Figure 2.3 Pipe specimens for ASTM D2688 [CERL—3/4-in (1.91-cm)]. (1) 3/4-in (1.91-cm) PVC service line; (2) 1-in (2.54-cm) union, PVC, socket type, Schedule 80; (3) 1-in 2.54-cm) × 10-in (25.4-cm) pipe nipple, PVC, Schedule 40; (4) 1-in (2.54-cm) × 3/4-in (1.91-cm) reducing bushing, PVC, socket; (5) 3/4-in (1.91-cm) PVC spacer, Schedule 40, OD reduced 0.015 in (0.038 cm) if needed, reducing union ID to pipe ID to hold specimen in place; (6) O-ring; (7) corrosion specimens, 3/4 in (1.91 cm), OD reduced 0.030 in (0.076 cm), machined from Schedule 40 galvanized steel or steel pipe; (8) specimen separator, 3/4 in (1.91 cm), PVC, Schedule 40, OD reduced 0.015 in (0.038 cm), length is adjusted to provide flush fit with inner faces of union. (*ASTM Book of Standards, vol. 11.01, sec. 11, D2688, American Society for Testing and Materials, Philadelphia, Pa., 1991.*)

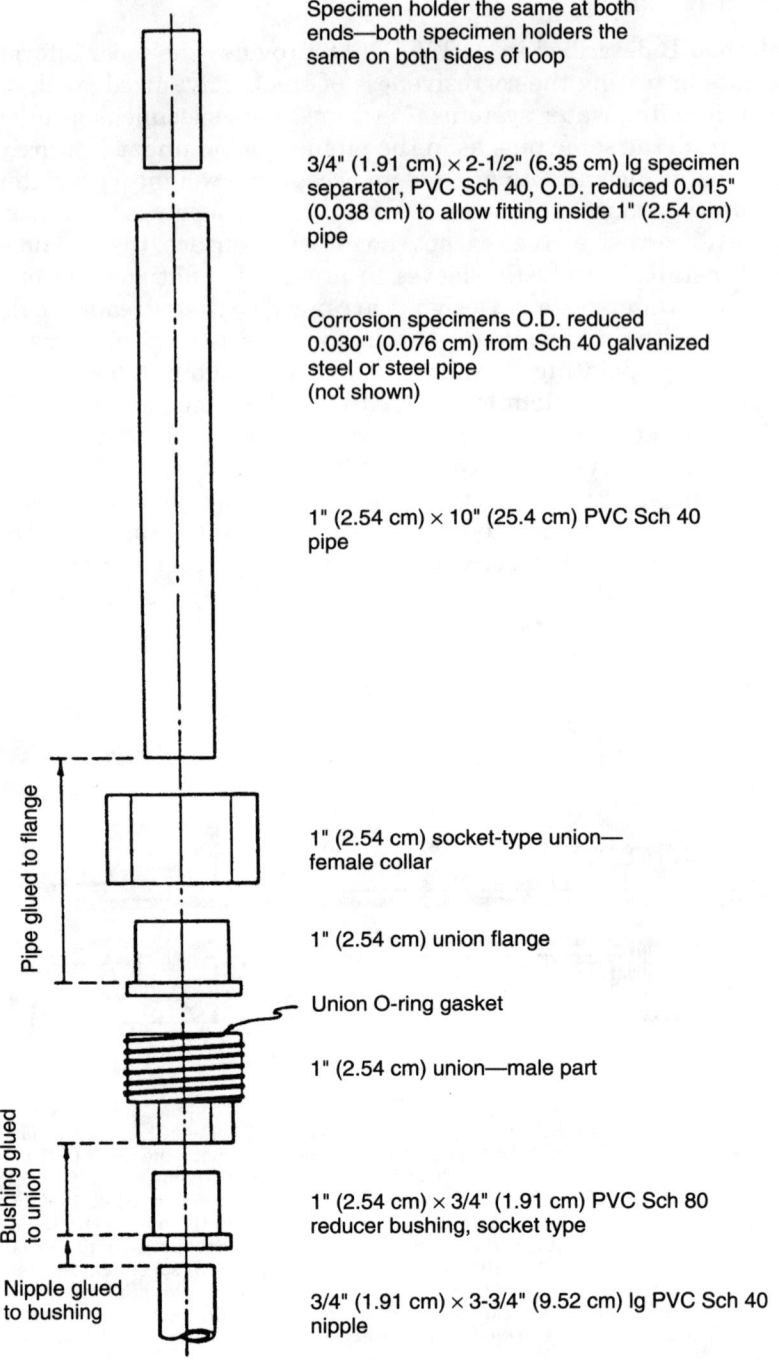

Figure 2.4 CERL pipe loop system—specimen holder details. (*ASTM Book of Standards*, vol. 11.01, sec. 11, D2688, American Society for Testing and Materials, Philadelphia, Pa., 1991.)

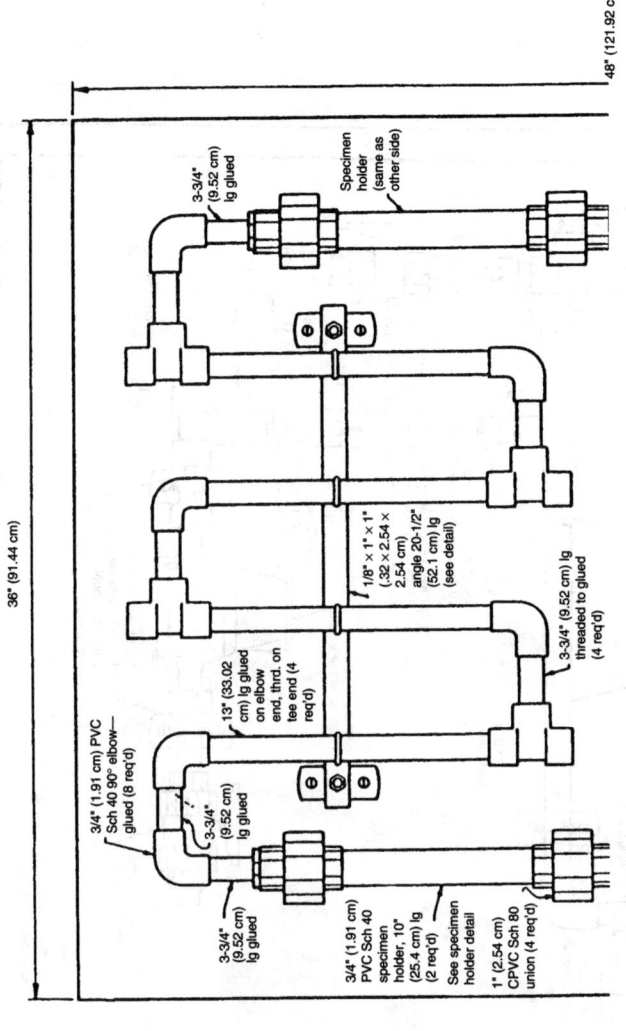

Figure 2.5 CERL pipe loop system—assembly drawing. (*ASTM Book of Standards*, vol. 11.01, sec. 11, D2688, American Society for Testing and Materials, Philadelphia, Pa., 1991.)

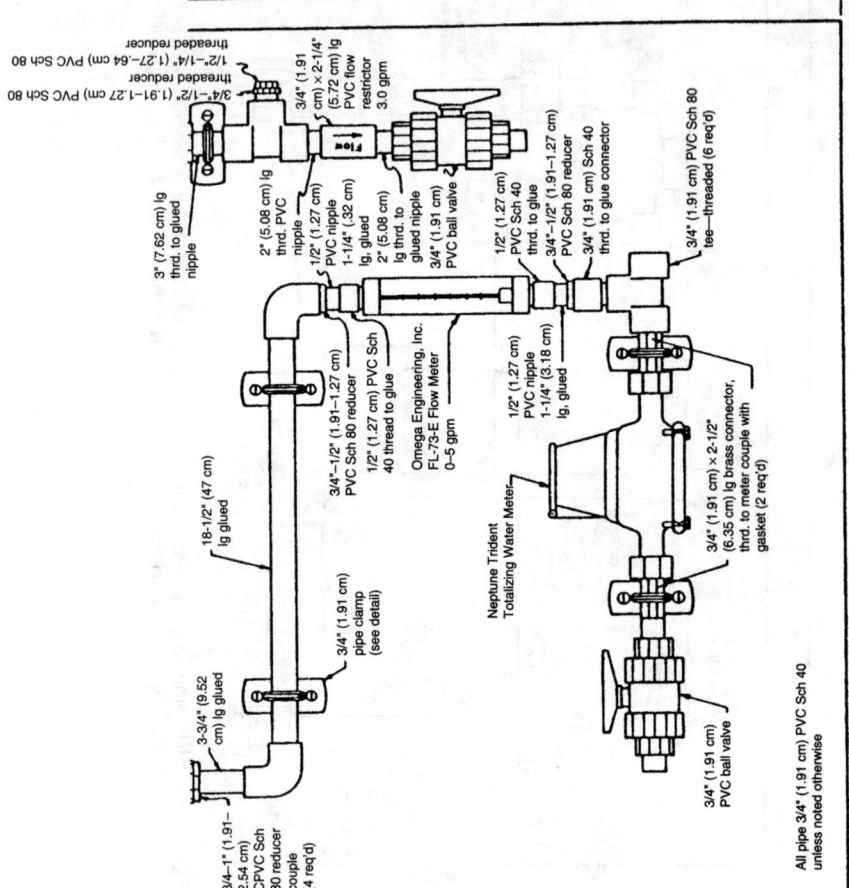

Figure 2.5 (*Continued*)

Corrosion of Galvanized Steel in Building Water Systems

Galvanized pipe is one of the oldest and most common plumbing materials used for domestic piping systems, although copper has now generally replaced it. The zinc coating of steel reduces the corrosion of steel by preventing water contact with the steel and by serving as the anode, which is corroded in preference to the cathode, steel. In this latter case, zinc protects the steel at holidays in the zinc coating. When zinc corrosion products, such as zinc basic carbonate, coat the zinc, its sacrificial property may be partially lost, and pitting of the galvanized steel may occur unless it is galvanically protected by other coatings, such as calcium carbonate. Numerous factors, such as the thickness and porosity of the zinc layer, its adhesion to steel, and the nature of the oxidation layer on the zinc surface, have a distinct bearing on the effectiveness of the zinc in protecting steel from corrosion.

On the basis of ASTM A-120, galvanized pipe is specified to have a minimum weight of 1.6 oz/ft^2 of zinc on steel and an average weight of 1.8 oz/ft^2 or more. Considerable variability in the weight of zinc per square foot is shown in pipe manufactured in the different countries[11] and supplied in the United States; however, the pipe may be expected to meet the minimum specification but may not always meet the specifications on uniformity of coating. Attention is directed toward keeping the cadmium below 0.01 to 0.2 percent and lead below 0.8 to 1.4 percent; cadmium and lead are observed as contaminants in the galvanized coating. There are a number of ASTM tests for evaluating the quality of galvanized steel, namely A90-81, A239-73, and A123-78:9.2, which are used for determining the weight of the zinc coating, the uniformity of coating, and the adherence of the coating.

Corrosion of Galvanized Piping by Water

There are five modes of failure with which to be concerned in the corrosion of galvanized pipe by water, namely:

1. General, or uniform, corrosion
2. Release of metals into solution
3. Pitting and tuberculation
4. Galvanic, or copper-induced, corrosion
5. Concentration-cell corrosion

In the case of the first—general, or uniform, corrosion—loss of the zinc for protecting the steel may lead to serious corrosion by pitting and tuberculation. One example of this would be a high-pH (>9.5) water causing increased dissolution of zinc into soluble sodium zincate, as occurs in high-pH lime-softened municipal waters and at a somewhat lower pH in galvanized cooling towers. In waters of low hardness (20 to 80 mg/L as $CaCO_3$) and low alkalinity (10 to 60 mg/L as $CaCO_3$), galvanized piping is subject to pitting-type corrosion because of a lack of adequate calcium bicarbonate to form a protective film of calcium or zinc carbonate.[12] In the presence of high chloride plus sulfate (200 to 1,000 mg/L), it is even more subject to pitting-type corrosion.[12-16]

Waters (including deionized water) of pH below 7.0 have been shown to appreciably corrode the zinc in galvanized steel and thus not provide the necessary cathodic protection of underlying steel. Galvanized pipe is also subject to corrosion at the pipe threads because much of the galvanizing is removed during the threading operation, leaving bare steel with lower metal thickness and inadequate zinc to provide cathodic protection. This can also be considered an example of crevice corrosion, in which the area under the crevice becomes deficient in oxygen and the area around the crevice possesses adequate oxygen—thus establishing an electrochemical cell for fostering the corrosion process.

Investigations have disclosed that the zinc layer likely corrodes initially at a rapid rate, causes a rapid buildup of iron in the protective scale layer, and later corrodes less rapidly as the zinc-iron protective layer is formed. The most effective cathodic protection of steel by zinc is attained when zinc is an anode large in area protecting the smaller steel cathode. Tests conducted on galvanized steel in 12 U.S. water supplies show it to be less corroded than steel, though it is less of an advantage in some waters than in others. Generally, acceleration of corrosion of zinc occurs in waters of higher oxygen content (such as surface waters), as a result of enhanced depolarization occurring in cathode areas.[8,17-19]

In waters containing appreciable calcium bicarbonate, as is normal in many well waters, galvanized steel is quite resistant to corrosion; however, at a pH less than 8.0 and in the absence of silicates and dissolved oxygen, zinc may be expected to go into solution as zinc ions. Anodic dissolution of zinc forms a film of zinc basic carbonate at a pH above 8.0; however, a film of calcium carbonate may form at exposed iron areas and reduce corrosive tendencies. Such films are not as effective in the presence of high concentrations of chloride and sulfate (150 to 700 mg/L) and insufficient concentrations of calcium and alkalinity.

According to J. R. Myers,[20] factors causing increased corrosion of steel and galvanized steel are temperature and velocity increases and the chemical composition of the water (in particular, dissolved oxygen and carbon dioxide, sulfate, and chloride contents). Reduction of the corrosion rate occurs when the water contacting the metal contains adequate calcium, alkalinity, and silica. While normally, temperature increases tend to increase the corrosion rate, increased temperature may also cause increased calcium carbonate scale formation and consequently a lower corrosion rate. Velocity increases may normally cause increased corrosion rates, while stagnant conditions may generally cause increased galvanized-steel corrosion.

A correlation between the pitting coefficient and the Ryznar Index (RI) has been established (see Chap. 3 for a means of determining this index). Myers[20] has developed the following prediction equations for determining the pitting coefficient and projected pit depths from data collected by the U.S. Army Construction Engineering Research Laboratory:

$$\text{In cold water: } P_c = 0.0200 \times (RI - 7)$$

$$\text{In hot water } [\leq 135°F\ (57°C)]: P_c = 0.0261 \times (RI - 7)$$

where P_c = pitting coefficient and RI = Ryznar Index.

Calculations: RI is obtained from a particular analysis of a cold water:

Calcium (as $CaCO_3$), 100 mg/L

M alkalinity (as $CaCO_3$), 38 mg/L

TDS,* 432 mg/L; pH at site, 7.9; RI = 8.4

$$P_c = 0.0200 \times (8.4 - 7) = 0.028$$

The following equations can be used to determine the projected pit depth P after t years:

$$\text{Cold water: } P = 0.0200 \times (RI - 7) \times t^{0.33}, \text{ in}$$

$$\text{Hot water } [\leq 135°F\ (57°C)]: P = 0.0261 \times (RI - 7) \times t^{0.33}, \text{ in}$$

Calculations: Using the same cold-water analysis:

$$P = 0.0200 \times 1.4 \times t^{0.33}$$

where t = 1 year and P = 0.028 in

t = 5 years and P = 0.048 in

t = 20 years and P = 0.075 in

Then on the basis of 0.133 in wall thickness of 1-in pipe, this pipe should last an estimated 22 to 35 years.

It must be admitted that these predictions have been based on indi-

---

*TDS = total dissolved solids.

vidual cases and that general application of these formulations may not always be accurate. At least these values provide a numerical starting point from which an experienced consultant can proceed and consider the importance of many other factors, such as temperature and velocity. The calculation of other indexes may be more informative—some are more suited to waters of lower or higher TDS. Also, knowing the chlorine, the chemical oxygen demand (COD) (an approximate measure of the organic matter), the buffer capacity, etc., may be helpful in predicting the corrosion rate and arriving at decisions on:

1. Whether piping systems should be replaced or repaired
2. Whether more effective water treatment should be installed
3. Whether different (more corrosion-resistant and expensive) materials of construction should be installed

The Illinois State Water Survey[21] determined the corrosion rate of galvanized-steel inserts (ASTM D2688, Method B[5]) exposed for 1-year periods in cold-water distribution systems in different water supplies at 25 locations and found the following relationship with water quality using multiple regression techniques:

$$\text{Corrosion rate (mdd)} = 2.0865 + [0.0289 \times (\text{mg/L SO}_4)]$$
$$- \{0.00296 \times [\text{M alkalinity (as CaCO}_3)]\} - \{0.045$$
$$\times [\text{mg/L calcium (as Ca}^{2+})]\} + [0.105 \times E09 \times (H^+)]$$

The corrosion rate (mdd) is the corrosion in milligrams per square decimeter per day; multiply this result by 0.2 to convert to the more common expression of the corrosion rate in mils per year (mpy).

This equation was shown to have a correlation coefficient of .83 when based on water analysis data of the following ranges in composition:

Sulfate (as SO_4), 5–226 mg/L

M alkalinity (as $CaCO_3$), 40–426 mg/L

Calcium (as Ca^{2+}), 7–75 mg/L

pH, 7.1–9.8 in which $H^+ = 10^{-pH}$

This equation should be helpful in determining the expected corrosion rate of galvanized steel in waters of the above composition. For example, if the corrosion rate is calculated to be 0.012 mpy and it is assumed that in 5 years the penetration will be 0.06 in, it should be of concern that half the wall thickness (0.133 in) of 1-in (schedule 40) galvanized pipe is likely to have been penetrated.

Under certain conditions, the electrochemical potential between zinc and iron can reverse so that iron becomes the sacrificial metal. This has been observed at high temperatures of 140°F (60°C) or at high bicarbonate (>80 mg/L) and/or appreciable nitrate concentrations; however, increased chloride, sulfate, calcium, and silicate are reported to counteract this reversal of potential.

While polyphosphate may prove effective in preventing calcium carbonate scale, experience indicates that this inhibition may prevent effective corrosion inhibition[22] of galvanized steel by inhibiting the formation of protective calcium carbonate scale.

Low hardness waters [<50 mg/L (or 3 gpg) as $CaCO_3$] are generally aggressive to galvanized piping. In systems where there is continuous circulation of water, this effect is much more prevalent than in households where flow is not continuous. In the midwest where waters of hardness above 200 mg/L are common, ion exchangers (sodium zeolite softeners) generated with salt are commonly installed to prevent scale formation in domestic hot water piping. While this technique serves this purpose well, complete softening to near zero hardness results in water that is very aggressive to galvanized piping. The solution for correcting this corrosiveness is *to blend sufficient hard water with the completely softened water to provide a water of 80 to 120 mg/L hardness.* A bypass line including a 3/4-in globe valve around the softener to blend about 25% hard water into the domestic hot water supply should be installed to solve this corrosion and scale problem. The aggressiveness of waters of naturally low hardness content may also be reduced by proportional application of sufficient sodium silicate to provide 8 mg/L as silicate (SiO_2).[23,24]

Galvanic corrosion

Traces (>0.1 mg/L) of soluble copper as acquired in minimal corrosion of copper heat exchangers in recirculating hot water systems can cause pitting of galvanized steel. Soluble copper will deposit on zinc or steel, forming small galvanic cells that result in active pits and possible perforation of metal. Galvanic corrosion of zinc is accelerated by contact with copper or copper alloys, as is observed when galvanized piping is connected to brass valves. In 1978 in California, a new housing development experienced serious failure of galvanized piping due to soluble copper in the water supply.[25] The soluble copper, derived from copper sulfate applied in an upstream reservoir for algae control, deposited on surfaces of the galvanized pipe and acted as a cathodic depolarizer. Since copper is a more noble metal, it was natural for the copper to plate out on the zinc and cause the zinc to go into solution as zinc oxide. Once a protective scale layer of calcium or zinc carbonate accumulated on the new pipe, the deposition of copper lessened and the rate of pitting of the galvanized piping declined.

Design of Domestic Hot Water Systems

In the design of domestic water systems, one should not mix copper and galvanized piping, and, in particular, one should not install copper preceding galvanized steel. These precautions are taken in order to avoid the situation in which soluble copper deposits on galvanized steel and causes galvanic attack. The author has been involved in litigation in which the contractor was considered guilty of poor design when copper equipment and piping were installed preceding a galvanized piping recirculating system. Another bad example can be the installation of galvanized-steel piping in a new addition to a building where the present piping system is copper and the two systems are part of a common hot water heater and circulating system. To counteract the effect of copper-induced corrosion of galvanized steel, silicate treatment can be applied to provide a protective scale layer on the galvanized steel; this prevents the soluble copper from causing pitting corrosion.

Cathodic Protection[26,27]

The installation of properly designed cathodic protection systems consisting of sacrificial anodes in cold and hot water storage tanks and heaters has proved beneficial in controlling corrosion in these areas. Essentially, this technique involves either the installation of sacrificial anodes, such as magnesium or zinc, which are more chemically reactive than the steel components, or the application of an electric current by nonsacrificial anodes, such as graphite, to counterbalance the natural corrosion current between steel and water. The water held in cold and hot water tanks is usually stagnant, making chemical treatment generally ineffective.

In the design of a cathodic protection system, the designer must make sure that anodes are properly spaced and adequate to provide the necessary counteractive current and must arrange the anodes so that they cover the whole surface of the tank subject to corrosion. Paints and linings may also be applied so that the active steel surfaces are minimal and less current or fewer anodes are required for attaining effectiveness. While magnesium anodes are desirable in that they are more reactive than other anodes, often they cause the development of a hydrogen sulfide odor, apparently from reduction of sulfate in the water supply. Substitution of zinc anodes eliminates this problem, though zinc is not as reactive a metal as magnesium and therefore is less efficient. There are many cathodic protection firms that efficiently design and service these systems.

Consideration should be given to the installation of cathodic protection to control corrosion encountered or to be expected in cold and hot water storage tanks, as it is an effective and inexpensive method of control.

Electric Grounding

Investigations have revealed that electric grounding of water pipe[28] may cause serious corrosion, particularly if insulating unions have been installed on each side of the water meter. If the electric current passes through the water to reach the connecting pipe, severe corrosion will result where the current leaves the water pipe. Also, meter inspectors may suffer a serious shock when performing maintenance on the water meter.

Maintaining domestic hot water temperatures above 135°F (57°C) will cause an increase in the corrosion rate and should therefore be avoided.

Corrosion of Copper in Building Water Systems

Copper, being a noble metal, has excellent resistance to corrosion. Failures of copper in the water industry are rare occurrences, and investigation into varied and seemingly unrelated fields may be required to determine the cause of corrosion when it does occur. Mechanical problems, plumbing design and selection of materials, water composition and treatment, and temperature and velocity of use are some of the factors that may have to be investigated and adjusted. Maintaining the cuprous oxide film (cuprite) is important for inhibiting corrosion in chlorinated potable water systems, in addition to maintaining the velocity and temperature within prescribed limits to ensure expected metallic integrity.

Copper first came into use as a plumbing material after World War II. It has replaced alternative materials up to 2-in pipe size owing to its superiority in corrosion resistance and ease of installation. The Copper Development Association reports that 80 percent of all tubing (equivalent to 500 million feet per year) is now being installed in water service and distribution systems. However, plastic piping in the form of polyvinyl chloride (PVC), polybutylene (PB), acylonitrile-butadiene-styrene (ABS), and/or chlorinated polyvinyl chloride (CPVC) is now offering a strong challenge to copper for installation in potable water systems.

Copper still serves as the more generally recommended material of construction for building water systems. It is a much more corrosion-resistant metal than galvanized steel in the areas where waters of lower hardness (<100 mg/L) predominate, as in the eastern and northwestern United States. While galvanized steel proves satisfactory for use in the middle western waters of high bicarbonate alkalinity, copper is superior in areas using surface water (high dissolved oxygen content) and in areas in which the chloride and sulfate content exceeds 150 mg/L (about 600 µS/cm conductivity).

The treatment methods specified in Table 2.2 were necessary to control scale and corrosion in water supplies used by Illinois state facilities. Both galvanized-steel and copper piping were observed to provide proper service in type A cold and hot waters [140°F (60°C)][29] that were derived from surface sources and were of moderate hardness (<150 mg/L), with low chloride and sulfate contents [($Cl^-+SO_4^{2-}$) <80 mg/L]. Treatment adjustments in the form of increasing the alkalinity to 50 to 100 mg/L and the pH to 8.3 to 9.0 were found necessary, with care taken to avoid scaling tendencies, as calculated from the Langelier Saturation Index (see Chap. 3).

Copper was observed to be the better choice for similar (type B) waters with a higher chloride and sulfate content (80 to 200 mg/L), which were preferably treated to provide a pH near 9.0.

As shown in Table 2.2, both copper and galvanized steel were observed to provide good service in nonaerated waters (type C) [high hardness (150 to 500 mg/L) and alkalinity (150 mg/L) and moderate chloride and sulfate content (<150 mg/L)], which were blended with zeolite-softened water to yield 60 to 100 mg/L hardness to inhibit corrosion and scaling.[29]

Copper was preferred for type D waters, which were high both in hardness (150 to 500 mg/L) and in chloride and sulfate content (<700 mg/L). These waters were blended with zeolite-softened water to provide 60 to 90 mg/L hardness to prevent scaling, and they were treated

TABLE 2.2 Treatment Required for Different Types of Water

Water type	Source	Hardness, mg/L	M alkalinity (as $CaCO_3$), mg/L	Chloride + sulfate ($Cl^- + SO_4^{2-}$), mg/L	Treatment required
A	Surface	60–150		<80	Raise alkalinity to 50–100 mg/L Raise pH to 8.3–9.0
B	Surface	60–150		80–200	Same as above
C	Nonaerated	150–500	150	<150	Soften to 60–90 mg/L hardness
D	Nonaerated	150–500		<700	Soften to 60–90 mg/L hardness Add silicate to 8 mg/L SiO_2 Adjust pH to 8.0–8.4

SOURCE: Illinois State Water Survey, Champaign.

with liquid sodium silicate [8 mg/L soluble silica (SiO_2)] and caustic soda (pH 8.0 to 8.4).[8,23,24]

Deficiencies of Copper as a Corrosion-Resistant Metal

What are its deficiencies? Free chlorine above 1 mg/L and chloramine (about 2 mg/L) are aggressive to copper, so treatment plant operators are cautioned to avoid overtreatment with chlorine and chloramine.[12,17,30,31] At an Illinois institution, application of 8 mg/L sodium silicate (as SiO_2) and/or a pH increase to 8.0 by caustic soda feed corrected a copper corrosion problem due to high chlorine (above 2 mg/L) in the water supplied by a municipality. Chloramine treatment, which involves either natural ammonia content or the application of ammonia before chlorination, also increases the corrosion of copper as a result of the complexing capabilities of ammonia.

Recently, when the author was visiting New Zealand and Australia and recalling their pitting problems with copper some ten to twenty years ago, it was learned that the pitting largely resulted from allowing new buildings to remain unoccupied for many months and allowing the systems to remain filled with stagnant water during those periods.[32] Apparently, the oxygen necessary to maintain a viable protective copper oxide film was used up during those stagnant periods, and as a result, the copper metal became subject to pitting corrosion. This was also experienced in California where new town house complexes unoccupied for several months developed copper pitting and leaks. A correlation has been established between the initial stagnation time and the development of leaks. On the basis of these experiences, a maximum period of 2 months has been recommended as the limit for allowing newly constructed houses to remain unoccupied without periodic flushing.[25,32,33] Periodic flushing corrects the problem; in particular, it removes debris and membranes under which pits may develop during stagnant conditions.

Copper is a softer metal than steel and therefore is subject to erosion-corrosion, particularly in the presence of carbon dioxide (low pH) at domestic hot water temperatures. For domestic hot water temperatures [<135°F (57°C)], tubing size should be specified to be large enough so that the maximum velocity is less than 4 ft/s (1.22 m/s), which is equivalent to 10-gpm flow in 1-in nominal size type K tubing. Sizing to 8 ft/s is considered satisfactory for cold water. In order to be assured that copper tubing of the correct size is being installed for the specified flow conditions, a Copper Tube Sizing Calculator should be obtained from the Copper Development Association, Inc. (260 Madison Avenue, New York, NY 10016). Erosion-corrosion results from the abrasive effects of high velocity (>4 ft/s) and is characterized by rounded holes and grooves in localized areas where the velocity and turbulence may be especially high. This may be observed particularly in areas of the piping system where there is a change in direction of the water flow (Fig. 2.2).

At Michigan State University,[34] new dormitories supplied with completely softened water flowing in undersized type L copper piping suffered severe erosion-corrosion and multiple leakage within a short period. The pitting observed was typical of erosion-corrosion: elongated pits or grooves, which resulted eventually in perforation. The solution was to blend sufficient hard water with the softened water to provide a less corrosive water of at least 60 mg/L hardness and to size piping so that a maximum velocity of 1 m/s (4 ft/s) (equivalent to 10-gpm flow in 1-in copper piping) would not be exceeded.

Impingement attack, a form of erosion-corrosion, has been observed in the faucet seats of valves contacted by water containing 2 to 3 mg/L chloramine.[7] Impingement has also been observed in power plant condensers where large air bubbles impinge on the inlet end of the tubes and break into small bubbles, causing disruption of the protective oxide film. Too high velocity or suspended matter (like sand) can also cause this.

In the structure of the cuprous oxide film, known as the primary protection against copper corrosion, cuprous ions may be missing on the metal surface, and so there are points where positively charged colloids, such as FeOOH, may deposit. This explains the effect of $FeSO_4$ as a corrosion inhibitor for copper. Malachite $[CuCO_3Cu(OH)_2]$ also provides a protective film, and passified pits may become active again if this film is disturbed.

Galvanic corrosion is also involved, since solder and steel or galvanized steel is anodic to copper and therefore provides cathodic protection of copper when close by. Large cathode areas (such as copper tubing) and small anode areas (such as solder) provide effective cathodic protection of the copper. Since solder, the anode, is corroded rather than the copper cathode, it should still be recognized that the copper is an important part of the overall corrosion cell.

While underground corrosion of copper piping is generally not a problem, there is evidence of such corrosion occurring where soils are particularly corrosive. Soil analyses should be conducted in cases of external corrosion, as soils differ in their corrosiveness, which depends on their wetness, aeration, corrosive contents, surface drainage, thermogalvanic cell possibilities, and stray currents (even from other cathodic protection systems).

Corrosion of copper from cellulosic insulation[35] has been experienced in cases of wet insulation contaminated with carbon, sulfur, and ammonium sulfate.

Pitting of Copper

Pitting was first thought to be caused by irregular carbon depositions from drawing oils used in manufacturing; however, the completeness and permanence of the cuprite film (a smooth, shiny scale) is now considered of most importance. Pits that remain active require more corrosion current, but as the cathode areas become covered, weak pits die, and if a higher pH is maintained, piping life as influenced by pitting may be extended to 10 to 40 years.

Pitting in cold water is the most prevalent incidence of corrosion of copper, followed by erosion-corrosion. Pitting corrosion is observed in waters that have a pH below 7.8 (high in carbon dioxide) and that contain more than 17 mg/L sulfate and a sulfate to chloride ratio of about 3:1. This cold water is usually a well water, containing more than 5 mg/L carbon dioxide and dissolved oxygen above 4 mg/L. The pitting is usually observed in horizontal lines, where gravity holds copper salts in the pits, and is more prevalent in new installations exposed to initial unfavorable water quality, but the pitting may be corrected after 3 to 4 years by the buildup of a protective oxide film.[36-39]

Copper is also subject to pitting in soft waters (hardness <60 mg/L, as $CaCO_3$) of low pH (<6.5) and in new pipe on which a protective oxide film has not yet developed.

Evidence of corrosion may be found in the green color of the water, an unpleasant taste, and green staining of fixtures. Observations indicate that a Cu_2O membrane forms over the pit containing $CuCl_2$ and Cu_2O crystals, which results in the formation of a bipolar electrode and in a reaction with water constituents. Precipitates of calcium carbonate ($CaCO_3$) and malachite [$CuCO_3Cu(OH)_2$] may then form. Such pitting usually develops on the bottom side (not vertical) of piping where debris collects and concentration cells develop. Periodic flushing to remove such accumulations is helpful in preventing this type of corrosion.

According to J. R. Myers,[40] the pit depth of copper tubing exposed to a cold aggressive water of the following analysis develops according to the following equation:

pH of 7.0–7.7; dissolved oxygen, 3 mg/L or above

Carbon dioxide, 15 mg/L or above

Chloride (Cl^-), 15 mg/L or above

Sulfate (SO_4^{2-}), 15 mg/L or above

$$P = 0.040t^{1/3}$$

where P is the pit depth in inches and t is the time in years. When this equation is used, calculations reveal that in 6 months the pit depth would be 0.032 in; in 1 year, 0.035 in; and in 3 years, 0.058 in—thus approaching the perforation of 1-in type K copper tubing having a wall thickness of 0.065 in.

In hot waters over 130°F (54°C) and containing

>0.1 mg/L aluminum

bicarbonate <75 mg/L

pH <7.6

a bicarbonate-sulfate ratio <1.5

the pitting attack can be predicted from the equation:

$$P = 0.0148t^{0.5}$$

With this equation, calculations reveal a 0.033-in penetration in 5 years, and in 20 years, 0.066 in should be expected.

Other important and appreciable factors affecting the corrosion rate of copper in building water systems are as follows:

1. The use of excessive and aggressive solder flux in joining the copper tubing.
2. Poor quality of workmanship involved in reaming and soldering connections.
3. Temperatures above 140°F (60°C).
4. The presence of manganese (>0.05 mg/L) or sulfide (>approximately 0.1 mg/L) in the water supply, which increases the corrosive tendencies.
5. The deposition of iron oxide, suspended matter, and other insoluble matter. These may form on the surface of copper-bearing metals and encourage the development of concentration cells and pitting corrosion.
6. Appreciable ammonia in the water supply, which may cause stress-corrosion cracking of copper-bearing metals.

Dr. Myers[40] has developed prediction equations for practically all of these variables.

The Illinois State Water Survey determined the corrosion rate for copper tubing (ASTM D2688 Method B)[5] inserts exposed to different cold distribution waters at 21 locations over a 1-year period[21] and found the following relationship with water quality:

Corrosion rate (mdd) = 2.993 − {0.03084 × [mg/L carbon dioxide

(as CO_2)]} + [0.001857 × (mg/L TDS)] − [0.3268 × (pH)]

This corrosion rate is presented in mdd, milligrams per square decimeter per day; so it is necessary to multiply the above result by 0.16 to present results in the more common expression of mils per year (mpy).

This relationship was shown to have a correlation coefficient of .87 and was obtained from waters of the following composition ranges:

TDS, 115–1,312 mg/L

CO_2 (as CO_2), 0.0 – 27 mg/L

pH, 7.1–9.7

The equation should be helpful in estimating the corrosion rate for copper tubing in waters of the above composition ranges. For example, a corrosion rate of 0.006 mpy might be considered serious, for in 5 years this could be 0.030 in, or nearly half the wall thickness of 0.065 in for 1-in copper tubing.

Extensive studies performed by the Illinois State Water Survey for the U.S. Environmental Protection Agency in 1987[19] revealed further information concerning the corrosion of copper tubing and the influence of water quality and plumbing materials on the metal concentrations in household drinking water systems.

Flux used to aid in soldering fittings installed in copper tubing systems has been observed to cause increased corrosion of copper tubing. Excess flux is sometimes applied and is not generally rinsed adequately out of the system during normal water usage. Ammonium chloride flux in particular has been reported to cause pitting of copper. CERL (U.S. Government Construction Engineering Research Laboratory)[41] has reported on an improved technique for rinsing the piping system with hot water [140 to 150°F (60 to 66°C)], a much better solvent for flux residues than cold water.

Water Treatment Methods for Inhibiting Copper Corrosion[42]

Application of 4 to 8 mg/L of Item 32 liquid sodium silicate[23] (41° Baumé, alkali-silicate ratio of 1:3.22) and sufficient caustic soda to maintain a pH of 8.0 is specified to inhibit pitting and general corrosion of steel, galvanized-steel, copper, and copper-alloy piping. A sufficient amount of completely zeolite-softened water should also be blended with hard water to provide water of 60 to 120 mg/L hardness (as $CaCO_3$).

Waters that are of low hardness and alkalinity (60 mg/L) and low pH (6.0 to 7.0) (>5 mg/L CO_2) and that show pitting propensity may be treated in the following way to correct pitting propensity:

1. Aeration to reduce carbon dioxide
2. Treatment with caustic soda, soda ash, liquid sodium silicate, or lime to raise the pH to 8.1 to 8.3
3. Passage through a neutralizing (limestone) filter
4. Treatment with limestone and carbon dioxide to increase the bicarbonate content

In Pinellas County, Florida,[43] pitting caused by a low-pH well water has been eliminated in the following ways:

1. Reducing dissolved oxygen and maintaining it at about 1 mg/L by construction of a 10-million-gallon reservoir with surface ventilation (exhaust fans) to remove carbon dioxide and hydrogen sulfide
2. Installation of cold lime softening to provide water of 100 mg/L hardness and controlled application of caustic soda and phosphate inhibitors

Copper Concentrations

Copper concentrations in potable water are observed to be higher in standing samples (particularly overnight), and so *it is recommended that water be allowed to flow from the faucet until household lines are flushed and water has reached the colder temperature of the municipal distribution system before members of the household drink coffee or prepare food.* Thirty seconds is estimated as the required time for flow, and slow opening and closing of the valve to avoid turbulence is also suggested. Water from the municipal distribution system is not exposed to the copper-bearing metals before entry into the household system and is therefore lower in copper content.

Copper concentrations will be found to be dependent on the area/volume ratio in flowing samples, and therefore 0.5-in tube may be expected to have double the concentration of a 1-in tube, but with static samples these should yield comparable results.

Soluble copper (1 to 5 ppm) is found in the first-draw samples in buildings supplied by many city water supplies, although the maximum contaminant level (MCL) for copper is 1.3 mg/L. It will be higher in waters of low pH and alkalinity, as the corrosiveness of a water depends entirely on its degree of saturation with the ions or molecules of the metal in contact. Raising the pH by alkali or silicate addition has been observed to reduce the amount of soluble copper.

Recently in a New York City office building, there was concern that copper was present to the extent of 3 to 5 mg/L in the low pH–low alkalinity drinking water supply; however, this red brass system (85 percent copper, 15 percent zinc) had never experienced a leak or corrosion failure in some 60 years. Flushing water fountains and faucets before use was recommended, although POU (point of use) treatment systems could have been installed at strategic sites. Installation of water treatment proportioning equipment for application of corrective treatment (raising the pH or adding a corrosion inhibitor) was deemed difficult and impractical owing to the many different water entry locations in the building.

Principal variables that affect the pickup of soluble copper from copper piping systems are as follows:

1. Water quality.
2. Materials in the system. Brass generally is more easily dissolved; copper and copper-nickel alloys are less easily dissolved.
3. Temperature and velocity.
4. Age of the plumbing system. In new systems dissolution is generally easier; older systems have protective oxide or scale layers, which prevent dissolution.
5. Length of time that water is in contact with metal.

Corrosion in Tanks

Installation of cathodic protection systems[26,27] has proved to be an effective means of reducing corrosion in metallic tanks, where stagnant conditions, dissimilar metals, and higher temperatures may increase corrosive tendencies. Coupling with a small sacrificial aluminum anode has been shown to prevent pitting in copper tanks and cylinders.

Corrosion of Brass

Brass is plagued by dezincification, particularly in the case of yellow brass (65 percent copper, 35 percent zinc). This is mainly caused by water containing high chloride (>40 mg/L), low temporary hardness (<100 mg/L, as $CaCO_3$) constituents, and chlorination.[38,44] Dezincification of brass occurs by one of two alternative mechanisms:

1. Dissolution of the copper and zinc, followed by redeposition of the copper

2. Selective dissolution of the zinc, leaving the copper behind as a spongy material

Season cracking of brass caused by simultaneous exposure to ammoniacal compounds and tensile stress (cold working or applied stress) is also experienced.

Red brass (85 percent copper, 15 percent zinc) behaves similarly to copper, is not subject to dezincification, and is generally chosen when corrosive environments are more severe. Admiralty brass and aluminum brass may be subject to stress corrosion cracking and associated pitting.

Case Histories

Analysis of a rust-colored deposit in a galvanized-steel fitting from a domestic hot water distribution line in a small institution revealed that the deposit was essentially brown iron oxide (Fe_2O_3) and showed an absence of calcium carbonate ($CaCO_3$). The question was this: Was the deposit caused by a lack of attention to the regeneration of the zeolite softener, or was a corrosion problem involved? The conclusion was that the deposition was caused by application of completely softened water to this galvanized-steel system for approximately 20 years. Some replacement of the galvanized steel with copper had already been performed, and so the presence of soluble copper causing possible mini-corrosion cells in the galvanized-steel system had probably also been a factor in the deterioration of the galvanized steel. Replacement of the entire system with copper was recommended, as calculations based on the analysis of this particular hard water indicated that copper was relatively corrosion-resistant in contact with blended hard water of approximately 100 mg/L hardness content. Calculations of various indexes using computer programs listed in Ref. 45 and App. D were helpful in reaching a solution to this corrosion problem.

In other case histories, the installation of small-sized copper tubing in hot water systems resulted in flows exceeding 1 m/s and caused erosion-corrosion and penetration of the tubing.

References

1. *Economics of Internal Corrosion Control,* American Water Works Association Research Foundation, Denver, 1989.
2. H. H. Uhlig and R. W. Revie, *Corrosion and Corrosion Control,* 3d ed., J. Wiley & Sons, New York, 1985.
3. K. R. Trethewey and J. Chamberlain, *Corrosion for Students of Science and Engineering,* Longman Scientific & Technical (John Wiley & Sons, Inc.), New York, 1988.
4. R. J. Landrum, *Fundamentals of Designing for Corrosion Control,* National Association of Corrosion Engineers, Houston, 1989.

5. *Test Methods for Corrosivity of Water in Absence of Heat Transfer (Weight Loss Methods)*, D2688-90, *1991 Annual Book of ASTM Standards,* vol. 11.01, American Society for Testing and Materials, Philadelphia, p. 218; also *Test Methods for Corrosivity of Water in Absence of Heat Transfer (Electrical Methods)*, D2776-79, p. 207.
6. C. P. Dillon, *Forms of Corrosion Recognition and Prevention,* National Association of Corrosion Engineers, Houston, 1982.
7. T. E. Larson, R. M. King, and L. M. Henley, "Corrosion of Brass by Chloramine," *Journal of the American Water Works Association,* January 1956, pp. 84–88.
8. T. E. Larson, *Corrosion by Domestic Waters,* Bulletin 59, Illinois State Water Survey, Urbana, 1975.
9. V. L. Snoeyink and D. Jenkins, *Water Chemistry,* John Wiley & Sons, Inc., New York, 1980.
10. J. M. Montgomery Consulting Engineers, Inc., *Water Treatment Principles and Design,* John Wiley & Sons, Inc., New York, 1985.
11. K. P. Fox, C. H. Tate, and E. Bowers, "The Interior Surfaces of Galvanized Pipe: A Potential Factor in Corrosion Resistance," *Journal of the American Water Works Association,* February 1983, p. 84.
12. R. W. Lane and C. H. Neff, "Materials Selection for Piping in Chemically Treated Systems," *Materials Performance,* vol. 8, no. 2, 1969, pp. 27–30.
13. R. A. Pisigan, Jr., and J. E. Singley, "Effects of Water Quality Parameters on the Corrosion of Galvanized Pipe," *Journal of the American Water Works Association,* November 1985, p. 76.
14. T. R. Camp, "Water and Its Impurities," Reinhold, N.Y., 1963.
15. I. G. Thompson, "Galvanized Steel Pipe in Potable Water Systems—The Corrosion Problem," Second Annual International Lead Zinc Research Organization (ILZRO) Galvanizing Seminar, St. Louis, June 1976.
16. D. J. Hubbard and C. E. Shanahan, "Corrosion of Zinc and Steel in Dilute Aqueous Solutions," *British Corrosion Journal,* vol. 8, 1973, p. 271.
17. R. A. Pisigan and J. E. Singley, "Influence of Buffer Capacity, Chlorine Residual and Flow Rate on Corrosion of Mild Steel and Copper," *Journal of the American Water Works Association,* February 1987, pp. 62–70.
18. *Plumbing Materials and Drinking Water Quality: Proceedings of a Seminar,* EPA/600/9-85/007, U.S. Environmental Protection Agency, Cincinnati, May 16–17, 1984.
19. C. Neff, M. R. Schock, and J. I. Marden, *Relationships between Water Quality and Corrosion of Plumbing Materials in Buildings,* vol. 1, *Galvanized Steel and Copper Plumbing Systems,* Illinois State Water Survey Contract Report 416-1, March 1987.
20. J. R. Myers, "Prediction Equations for the Pitting Corrosion of Steel and Galvanized Steel by Domestic Waters: The First Iteration," prepared for U.S. Government Construction Engineering Research Laboratory (CERL), December 1987.
21. C. H. Neff, F. W. Sollo, and R. W. Lane, unpublished investigation, 1975.
22. T. Holm and M. Shock, "Potential Effects of Polyphosphate Products on Lead Solubility in Plumbing Systems," *Journal of the American Water Works Association,* July 1991, p. 76.
23. R. W. Lane, T. E. Larson, C. H. Neff, and S. W. Schilsky, "Silicate Treatment Inhibits Corrosion of Galvanized Steel and Copper Alloys," *Materials Performance,* vol. 12, no. 4, 1973, p. 32.
24. R. W. Lane, T. E. Larson, and S. W. Schilsky, "The Effect of pH on the Silicate Treatment of Hot Water in Galvanized Piping," Illinois State Water Survey Reprint No. 367, Champaign, 1977.
25. K. P. Fox, C. H. Tate, G. P. Treweek, R. R. Trussell, A. E. Bowers, M. J. McGuire, and D. D. Newkirk, *Copper-Induced Corrosion of Galvanized Steel Pipe,* EPA/600/S2-86/056, U.S. Environmental Protection Agency, September 1986.
26. L. P. Sudrabin, "Controlling Corrosion in Water Systems," *Chemical Engineering,* September 1956.
27. J. R. Myers and M. F. OBrecht, "Corrosion Control for Potable Water Storage Tanks," *Heating Piping Air Conditioning,* January 1977, p. 61.

28. A. Horton, "Corrosion Effects of Electric Grounding of Water Pipe," *Corrosion/91*, Paper 519, Cincinnati, March 1991.
29. R. W. Lane and C. H. Neff, "Materials Selection for Piping in Chemically Treated Systems," *Materials Protection,* vol. 8, no. 2, 1969, pp. 27–30.
30. D. Atlas, J. Coombs, and O. T. Zajicek, "The Corrosion of Copper by Chlorinated Drinking Waters," *Water Research,* vol. 16, 1982, pp. 693–698.
31. H. Ingleson, A. M. Sage, and R. Wilkinson, Jr., "The Effect of Chlorination of Drinking Water on Brass Fittings," *Institute of Water Engineers,* vol. 3, 1949, pp. 81–90.
32. G. G. Page, personal communication, Auckland, New Zealand, 1989.
33. D. M. Waters, Waters Consultant, "Copper Corrosion in Domestic Water Services," San Diego, 1974.
34. "How Temperature, Velocity of Potable Water Affect Corrosion of Copper and Its Alloys," *Heating, Piping and Air Conditioning,* January 1960–April 1961.
35. *Corrosive Attack of Copper Water Tube by Cellulosic Insulation,* Copper Development Association, January 1978.
36. A. Cohen and J. R. Myers, "Mitigating Copper Pitting through Water Treatment," *Journal of the American Water Works Association,* February 1987, pp. 58–61.
37. "Pitting in Copper Water Tubing," *Journal of the American Water Works Association,* October 1985, pp. 70–73.
38. E. Mattsson, "Corrosion of Copper and Brass: Practical Experience in Relation to Basic Data," *British Corrosion Journal,* vol. 15, no. 1, 1980, pp. 6–13.
39. V. F. Lucey, "Mechanism of Pitting Corrosion of Copper in Supply Waters," *British Corrosion Journal,* vol. 2, no. 9, 1967, pp. 175–185.
40. Personal communications from Dr. J. R. Myers, corrosion/metallurgist consultant, 1991.
41. V. L. Van Blaricum, O. S. Marshall, R. H. Knoll, and V. F. Hock, "A Rehabilitation Technique for Soldering Flux-Induced Corrosion in Copper Potable Water Piping Systems," *Corrosion/91,* Paper 339, Cincinnati, March 1991.
42. H. Cruse and R. Pomeroy, Jr., "Corrosion of Copper Pipes," *Journal of the American Water Works Association,* August 1974, p. 479.
43. "Copper Corrosion Problem in Pinellas County Water System, Florida," unidentified paper, 1979.
44. M. R. Shock and C. H. Neff, "Trace Metal Contamination from Brass Fittings," *Journal of the American Water Works Association,* November 1988, pp. 47–56.
45. R. W. Lane, "Using Computer Programs to Control Corrosion and Scale in Water Systems," *Corrosion/88,* Paper 110, National Association of Corrosion Engineers, St. Louis, March 1988.

Chapter 3

Potable Water

Potable water exhibiting off-flavor, odors, turbidity, and discoloration is of concern, particularly since it could be contaminated from a cross connection with a nonpotable water system in the building. Water supplies originating from municipal water treatment systems are properly disinfected to render the water disease-free, but occasionally in the case of a surface water supply, some turbidity may enter the distribution system. Corrosion as well may occur in the piping of an aged city distribution system, causing off-flavor and discolored water, possibly accentuated at dead ends or by fire-oriented flow conditions. While most cities treat their supplies so that stable water (noncorrosive and non-scale-forming) is provided, others do not, and in the case of low hardness and low solids waters, as in the east and northwest, there may not be a simple or economical solution for providing the desired water quality or completely effective corrosion control. In such cases, installation of supplemental softening, filtration, or chemical treatment should be given consideration, provided these do not conflict with local ordinances.

Cold water, being at a lower temperature, does not generally require treatment because the chemical reactions associated with incidences of corrosion and scaling either do not proceed or are slow to proceed. This is particularly true if the water supply is derived from a municipally treated supply that is under the jurisdiction of the state Environmental Protection Agency (EPA). Softening is therefore not generally required for cold water used for drinking, flushing, sprinkling, etc.; however, the water should be iron-free, noncorrosive, and stable (as determined from the calculated value of the Langelier Saturation Index, which is discussed later in this chapter).

Criteria for Potable Water

The criteria for potable water were set forth by Congress in 1974 in the form of the National Safe Drinking Water Act, to which the Interim

Primary Drinking Water Regulations were added in June 1977. This act charged the EPA with responsibility for developing drinking water regulations to protect the public health.

Standards listing MCLs (maximum contaminant levels) were made effective in June 1977 and have since been amended—in 1979, 1982, 1986, 1989, 1990, and 1991. A substantial part of these later amendments pertains mainly to lead, copper, and organics. Since the organics limits are of chief concern to the water treatment plant and not generally to buildings, the rather lengthy list of such limits is not included in Table 3.1. The lead and copper rule is of primary concern to

TABLE 3.1 Federal Drinking Water Standards, Including Primary MCL and Secondary (Aesthetic) MCL Guideline Values

	MCL	
	Primary guideline (mandatory), mg/L	Secondary guideline (aesthetic) (goals), mg/L
Aluminum	0.02	
Arsenic	0.05	
Asbestos	No guideline value set	
Barium	No guideline value set	
Beryllium	No guideline value set	
Cadmium	0.005	
Chloride		250
Chlorobenzenes and chlorophenols	No guideline value set	
Chromium	0.05	
Color	True color units (NTU)15	
Copper	1.3*	
Cyanide	0.1	
Detergents	No foaming; no guidelines set	
Fluoride	1.5	
Hardness	No health-related guideline set	500 (as $CaCO_3$)
Hydrogen sulfide		Not detected by consumers
Iron		0.3
Lead	0.015†	
Manganese		0.05
Mercury	0.001	
Nickel	No guideline value set	
Nitrate	10 (as N) nitrite	
Oxygen—dissolved		No guideline value set
pH	6.5–8.5 units	
Selenium	0.01	
Silver	No guideline value set	
Sodium	No guideline value set	200
Solids—total dissolved		1,000
Sulfate		400
Taste and odor		Inoffensive to consumers
Temperature		No guideline value set
Turbidity	(nephelometric turbidity units) 5 (preferably <1)	
Zinc		5.0

TABLE 3.1 Federal Drinking Water Standards, Including Primary MCL and Secondary (Aesthetic) MCL Guideline Values (*Continued*)

Radioactive Constituents	
Gross alpha activity	0.1 Bq/L
Gross beta activity	1 Bq/L

Supplement: Maximum Contaminant Level, mg/L‡			
Inorganic chemicals:			
Antimony	0.006	Nickel	0.1
Beryllium	0.004	Thallium	0.002
Cyanide	0.2		
Volatile organic compounds:			
Dichloromethane	0.005		
1,2,4-Trichlorobenzene	0.07		
1,1,2-Trichloroethane	0.005		
Pesticides:			
Dlapon	0.2	Glyposate	0.7
Dinaseb	0.007	Oxamyl (Yydate)	0.2
Diquat	0.02	Picloram	0.5
Endothall	0.1	Simazine	0.004
Endrin	0.002		
Other organic compounds:			
Benzo(a)pyrene		0.00002	
Di(2-ethylhexyl)adipate		0.4	
Di(2-ethylhexyl)phthalate		0.006	
Hexachlorobenzene		0.001	
Hexachlorocyclopentadiene		0.05	
2,3,7,8-TCDD (dioxin)		3×10^{-8}	

*Copper action level is exceeded if concentration of copper in more than 10% of the samples collected during any monitoring period is greater than 1.3 mg/L.

†Lead action level is exceeded of concentration of lead in more than 10% of the samples collected during any monitoring period is greater than 0.015 mg/L.

‡On June 1, 1992, the U.S. Environmental Protection Agency set levels that are to go into effect for 23 more compounds. As this agency updates levels at various times, periodically checking with the agency directly is suggested.

SOURCE: U.S. Environmental Protection Agency.

building managers, homeowners, and occupants, as the levels specified are very low, and metallic contamination from lead and copper piping in these facilities has been found to be of serious concern and accordingly needs critical examination.

Lead Contamination

In 1986, the MCL (maximum contaminant level) for lead was set at 0.050 mg/L (50 ppb), and a ban was placed on the use of materials in pipe and fittings exceeding 8 percent lead and on the use of solder and flux exceeding 0.2 percent lead in any plumbing system supplying water for human consumption. As a result of this ordinance, tin-antimony and silver solder must now be substituted for tin-lead solder in joining copper in all new construction.[1] On the basis of the latest corrections to the lead and copper rule dated July 15, 1991, treatment technique requirements must be instrumented when the action levels for lead (0.015 mg/L) and copper (1.3 mg/L) are exceeded, as measured in the ninetieth percentile at the customer's tap. The actual MCLG (maximum contaminant level guideline) for lead is set at zero; this is supported by evidence of the high lead levels in the blood of children and by the abundance of evidence that lead can cause serious damage to the brain, kidneys, nervous system, and red blood cells. Treatment requirements consist of optimal corrosion control, source water treatment, public education, and lead service line replacement.

An aggressive municipal water supply may corrode galvanized steel or copper (and red and yellow brass) water distribution piping in a building, and the corrosion results in discolored, off-flavor, and turbid water, as well as drinking water exceeding the MCLs for lead and copper.[2,3] In these cases, it may be necessary to apply water treatment corrosion inhibitors in the form of silicate or zinc orthophosphate or alkaline chemicals recommended by a qualified water treatment specialist. Installation of a filter may be necessary to remove the turbidity of the city water or to remove corrosion products deposited or formed in the building distribution systems (deposition of corrosion products in the piping system can lead to underdeposit corrosion).

Chilled Potable Water

Water from drinking water coolers and central chilled drinking water systems installed in many large buildings should be free of off-flavors, and discoloration. There have been cases in which galvanized steel, copper, or brass piping (as listed in the ASHRAE 1988 *Equipment Handbook*[4]), was installed for this service, and unsatisfactory conditions involving serious off-flavors and discolored waters were reported. Such systems may have areas where there is little or no flow, with the result that this piping may corrode, particularly under stagnant conditions, and as a result, undesirable colored and off-flavored water containing corrosion products may be in evidence. Stainless steel is the proper material to install in such systems; then drinking water free of discoloration or turbidity can be provided.

Water cooler systems, consisting of a mechanically refrigerated assembly, differ from water chillers in that they dispense potable water rather than cooled water for air-conditioning purposes. They may include a bottle-type water cooler, a pressure-type water cooler, or a remote-type cooler.

There are many types of automatic ice makers,[5] but all are subject to the innumerable problems to be expected if the water supply is not of top quality with respect to scaling tendency, turbidity, or dissolved solids content (in excess of 400 mg/L). While demineralization is the best overall method of water conditioning, sodium cycle softening will be adequate to eliminate scaling problems unless the water exceeds 400 mg/L dissolved solids content; then cloudy or mushy ice may develop. Polyphosphates [such as Item 8 or 23 (see App. A)] may be used to reduce the scaling problem, but it may be preferable to contact a local water treatment company in order to arrive at the simplest solution.

Waters high in hardness and the Langelier calcium carbonate Saturation Index (LSI)* may cause scale formation and possible plugging of the piping, although this is not likely at the cold temperature of a chilled potable water system.

Chlorination

The flavor from chlorination is objectionable to some people, but personally noting this flavor assures me that the water is likely to have been adequately treated. Many cities now rely on chloramine treatment, a combination of ammonia and chlorine for disinfection. This method of treatment is preferred for surface water supplies, since chlorine reacts with the high organic content often associated with surface water supplies to form trihalomethanes. The U.S. Environmental Protection Agency (EPA) has fixed a limit of 0.1 mg/L on trihalomethanes, as research has indicated that these chemicals are carcinogenic. However, there is concern that depletion of chlorine residual under conditions of warmer temperatures and relatively stagnant flow may account for bacterial growth to the extent of two or three orders of magnitude in building plumbing systems.[6] The municipal water treatment plant has the responsibility of treating the water to make it stable and free of bacteria so that appreciable depletion of chlorine and bacterial contamination do not occur in the building water system.

Figure 3.1 shows that corrosion increases as the chloride/bicarbonate ratio increases. This initial work undoubtedly led to the concept of the Larson Index, described later in this chapter. Figure 3.1 also significantly points out the advantage of employing chloramine treatment rather than free chlorine in inhibiting corrosion of steel.

Chloramine (ammonia + chlorine) treatment may also be preferred because chloramine does not react with the organic content of surface water supplies; in addition, it has the advantage of being *more stable*

---

*See the section in this chapter entitled "Indexes for Estimating Corrosive and Scale Tendencies of a Water."

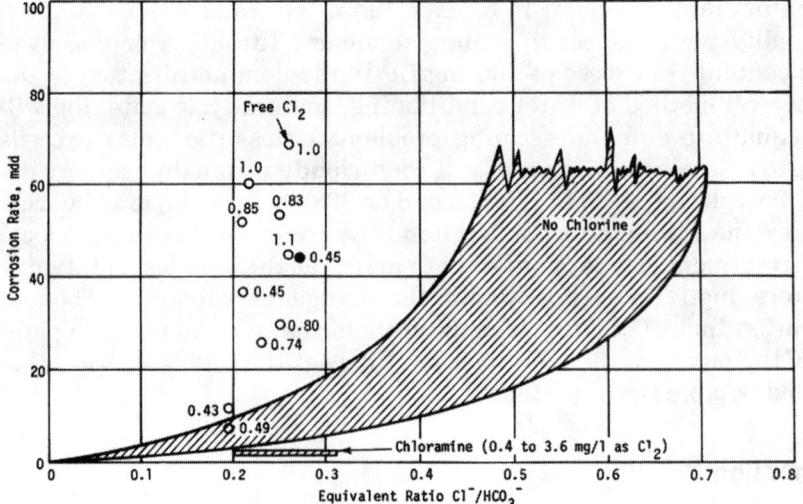

Figure 3.1 Relative corrosion rates at particular chloride/bicarbonate ratios with and without chlorine. (*Illinois State Water Survey, Champaign, Ill.*)

than chlorine, though it is less effective and slower in reaction as a disinfectant. However, a recent investigation[7] has disclosed that the plastic in gaskets and toilet reservoir floats is attacked by chloramine-treated water and requires the substitution of a more resistant plastic. Chloramines and un-ionized ammonia also have adverse effects on fish life in aquariums. The mechanism is similar to that experienced in the disease methemoglobinemia occurring in infants drinking water of high nitrate (>45 mg/L as NO_3) content. While dechlorination by sulfite is effective, ammonia removal by ion exchange, as well as proper pH control, is also usually required.

POU Treatment

Installation of point-of-use (POU) devices consisting of activated charcoal filters is a common solution to the off-flavor problems (even those attributed to chlorine) observed in potable water in residences. Such devices in larger sizes are also commonly installed in building systems that have circulating chilled potable water. The need to replace the filter cartridges periodically is often not given adequate attention. This can become of serious concern because bacterial growth in these cartridges can develop to the point where the water may develop off-flavors and be the cause of infectious diseases. Some of these devices include silver, which has bacteriostatic properties when applied correctly and in the proper form; however, to be continually effective, the silver must retain its active surface and dissolution characteristics and not become coated with films, which make it inactive. Recently, filtration devices that contain activated carbon in particle form and include periodic backflushing have been developed. These devices appear to correct the problem of bacterial growth areas developing in the bottom of the filter beds.

Bottled Water

Bottled water is regulated by the Food and Drug Administration (FDA) as a food under the Federal Food, Drug and Cosmetic Act.[8] The FDA quality standards are essentially the same as the latest EPA MCLs for drinking water, and although bottled mineral water is not covered, the FDA is considering a similar standard for it. The FDA's jurisdiction pertains to interstate commerce; intrastate shipments are the responsibility of the state. The International Bottled Water Association claims that its members' products are bottled in facilities regulated like food plants and are quality products. The FDA's capability to monitor these plants is limited, and since the product is considered a low public health problem, plant inspections are provided only every four years unless the firm is found to be in violation as determined by analysis or receipt of complaints. There has been some clamor to provide more effective enforcement, perhaps brought on by articles and by an investigation[9] in which 37 brands of *mineral* waters (9 domestic and 28 imported) from the Chicago and Pittsburgh markets were tested. The tests revealed that 24 (4 domestic) were out of line with U.S. drinking water standards. As a result of this investigation, it was suggested that drinking tap water might be better with regard to health, since this high degree of noncompliance would not be expected in community water supplies. To obtain more details on the quality of bottled water, read Chap. 6 of *The Drinking Water Book*.[10]

Corrosion Inhibitor Treatment of Potable Water

Corrosion inhibitors in drinking water systems are limited to USEPA accepted chemicals, such as silicate, polyphosphate, orthophosphate, zinc polyphosphate, and zinc orthophosphate. The dosage of these chemicals is generally limited to 10 mg/L.[11,12] The addition of 4 to 8 mg/L of sodium silicate has proved to be effective in reducing the corrosion of steel and copper.[13] Adding 2 to 4 mg/L of sodium polyphosphate (known as threshold treatment) has proved to be effective in preventing the deposition of calcium carbonate scale at normal domestic hot water temperature [135°F (57°C)]; however, a higher dosage (>20 mg/L) is required for inhibiting the corrosion of steel. Sodium polyphosphate is effective in treating (<1 mg/L as Fe) for the prevention of iron staining, discoloration, and iron-laden water in the dead ends of municipal distribution systems when applied at dosages of twice the iron concentration or more. Zinc polyphosphate and zinc orthophosphate are effective in reducing the corrosion of steel, copper, and lead at dosages of 0.5 to 1.0 mg/L zinc and at a pH of 7.5 to 8.0. Zinc orthophosphate is generally more effective than the polyphosphate.

Domestic Hot Water

Corrosion and scale formation may become more of a problem when water is heated for lavatory, bathing, laundering, and dishwasher uses. Keeping the temperature no higher than 135°F (57°C) will minimize corrosion and scale problems; however, the dishwasher requirement of 180°F (84°C) may necessitate the installation of a more corrosion-resistant material (such as stainless steel or a copper-nickel alloy), as well as water-softening equipment. In the case of moderately hard waters, it may be preferable to periodically clean the piping with an inhibited acid-type formulation.

Since domestic hot water systems are much more subject to corrosion and scale problems, it is not to be unexpected that plugging of piping from scale and corrosion product buildup, discoloration of water, inefficient heat transfer in heating and cooling equipment, and serious corrosion causing penetration of piping and leakage may occur in building water systems. Also, reduced pressure and flow, reduced capacity for heating water, and malfunctioning of control components, ice cube machines, and miscellaneous equipment can result from water turbidity, scale, and corrosion products. The life of the water system may be seriously effected by these water deficiencies.

Observing the appearance and color of the deposition causing flow interruptions or water turbidity is helpful in determining the necessary corrective treatment. Generally, a continuous layer of an off-white deposit is identified as calcium carbonate scale. This may be verified by observing that fizzing (carbon dioxide evolution) occurs when a drop of hydrochloric acid is applied to the deposit. This evidence of deposition of calcium carbonate indicates that a softener, such as a sodium ion exchange softener, should probably be installed to remove hardness from the water. These softeners (described fully in Chap. 7) remove hardness from the water by contact with the ion exchange resin and by the exchange of calcium and magnesium hardness for sodium in passage through the resin bed. They are relatively inexpensive and can be specified to include periodic automatic salt regeneration (initiated by a timer or by gallonage), requiring only infrequent addition of salt to the brine tank for regenerating the ion exchange resin.

The appearance of brown staining of porcelain lavatory surfaces, rust-colored suspended matter, hard tubercles of rust, and partial plugging of piping with rust may likely be the result of the corrosion of steel or galvanized steel in the system. This corrosion could have been caused by completely softened water (zero hardness) resulting from sodium ion exchange treatment. In this case, blending with hard water (about 15 to 30%) to provide 80 to 120 mg/L hardness (as $CaCO_3$) may be the solution to the corrosion problem. In case the water is of low hardness [<50 mg/L hardness (as $CaCO_3$)], application of a corrosion inhibitor such as sodium silicate or an alkaline chemical such as soda ash to raise the pH of the water (>8.0) may be required.[12,13] If the galvanizing or zinc that covers the steel has been gradually corroded away, the steel underneath corrodes, and the same appearances of rust and brown discoloration and tubercles are shown as with plain iron or steel.

Blue or green discoloration on lavatory surfaces or blue-colored water is likely the result of corrosion of copper or brass. This also indicates the need for application of a corrosion inhibitor such as sodium silicate or soda ash to raise the pH. It is not uncommon to observe both scaling and corrosion occurring in a system, in which case installation of a softener, partial blending back of part of the hard water, and application of a corrosion inhibitor may be required to correct the observed conditions.

Softening is generally not required if the hardness is below 130 mg/L, and only partial softening is required to provide water of 80 to 120 mg/L hardness when waters are higher in hardness than 130 mg/L. In homes, where hot water is not circulated, completely softened water has been observed to be somewhat less corrosive under these flow conditions. In the case of highly corrosive waters, characterized by high conductivity (>600 µS/cm), partial softening to 80 to 100 mg/L hardness plus pH adjustment with Item 32 liquid sodium silicate (see App. A) to obtain a pH of 8.0 or above (a slightly positive Langelier Saturation Index) are prescribed;[12,14] 14 to 28 mg/L (1.2 to 2.4 lb/10,000 gal) of Item 32 liquid sodium silicate is required to provide the necessary 4 to 8 mg/L silica (SiO_2) in hot water for corrosion inhibition. In waters of low hardness and alkalinity (20 mg/L), the protective scale layer formed by sodium silicate treatment is basic zinc pyrosilicate, but in waters of higher hardness (80 to 150 mg/L) and alkalinity (120 to 350 mg/L), the scale layer is basic zinc carbonate.

Keeping hot water temperatures below 135°F (57°C) is recommended to keep bacterial counts under control, and these temperatures should not go above 135°F so that corrosive tendencies are kept under control. This is particularly necessary in the case of galvanized-steel piping, since at the higher temperatures the steel in galvanized-steel piping is reported to become anodic to zinc (normally the sacrificial metal), and as a result, the piping may become seriously corroded.

Reference to TPC Publication[15] of the National Association of Corrosion Engineers (NACE) is recommended for detailed information on auxiliary equipment such as tanks, pumps, temperature controls, hot water generators, linings, expansion joints, faucets, and valves, which are installed in domestic hot water systems. In cases where magnesium anodes have been installed in hot water storage tanks to provide cathodic protection, it may be necessary to remove them to stop the hydrogen sulfide odor problem. Replacement with a properly designed zinc electrode system will provide the desired cathodic protection without the hydrogen sulfide problem.

The life expectancy of home water heaters will vary from possibly a few years to 20 or more years, depending on the corrosive tendency of the water supply. This can be determined by analyzing the water supply, determining the various water indexes, and applying the computer programs provided in App. D. In areas where the water supplies are more corrosive, the need for installation of anodes for application of cathodic protection will be more in evidence. This need is probably not as serious as in the past, because water heater manufacturers are apparently providing more satisfactory and complete protective coatings (no pinholes) in the manufacturing of tanks. The installation of partial zeolite softening will largely eliminate undesirable hardness sludge accumulations in the tanks. Keeping the water temperature near 135°F (57°C) should reduce bacterial growth and not cause the corrosion rate increases that are expected at higher temperatures.

Corrosion Testing

The installation of test nipples or corrosion testers, as described in Chap. 2, is recommended for monitoring the water system.[14] These enable the building operator to determine whether presently installed materials are proper or require replacement or whether the present water treatment should be changed to assure that the system will have a long life.

Corrosion of Copper and Its Role in Corrosion of Galvanized Steel

Copper is usually preferred to galvanized steel in most hot waters, although galvanized steel proves satisfactory in the high-bicarbonate-alkalinity waters in the midwest. Copper and galvanized-steel piping should not be mixed in a piping system, as galvanized steel is anodic to copper and may be sacrificially attacked when the two metals are adjacent to each other. The same or compatible materials should always be installed in piping systems in order to avoid galvanic corrosion. If any corrosion of copper occurs and soluble copper results, even as little as 0.5 mg/L may plate out on galvanized steel and cause the development of an electrolytic cell on the galvanized steel; subsequently, pitting will result.

The author has been involved in legal controversies that involved water flow that first contacted copper before the galvanized-steel piping; early failure of the galvanized steel occurred. As a result, the contractor, who performed this unorthodox installation, was held accountable for damages.

It is common practice to install brass valves in a galvanized pipe system and to experience later the failure of the galvanized pipe threads at the junction with the brass valve. This is an expected case of galvanic corrosion in which steel (galvanizing is removed in the threading operation) is the anode and the copper the cathode. A solution is to install an insulating coupling between the two metals. While the galvanic corrosion rate is affected by numerous factors, the dissolved solids (or conductivity) can be particularly important. A high conductivity (>600 µS/cm) may cause such failures within 5 years, while a lower conductivity and less aggressive water may not cause such failures for 25 years. Considering this, the builder must decide whether the benefit to be derived from installing insulated couplings between dissimilar metals in the building system is worth the additional cost.

Indexes for Estimating Corrosive and Scale Tendencies of a Water

For years, attempts have been made to predict the corrosiveness of a water from knowledge of the composition of the water; however, these attempts have only been partially successful. To date, the calculation and use of the Langelier calcium carbonate Saturation Index (a measure of the tendency of calcium carbonate to dissolve or deposit in a water system) of a water has provided the most information concerning the scaling and corrosive tendency of a water. Waters having a positive index traditionally have been assumed to be protective, while those having a negative index have been assumed to be nonprotective or corrosive to steel.

Langelier Saturation Index[16]

In 1936, W. F. Langelier came forth with the calcium carbonate Saturation Index, an evaluation of calcium carbonate saturation in water solution. About this time, knowledge developed concerning water chemistry equilibriums, and a practical method for determining pH became available. If the calcium [Ca^{2+} (as $CaCO_3$)], the total alkalinity [M alkalinity (as $CaCO_3$)], and temperature A and total dissolved solids B of a water supply are known, the pH of saturation pH_s of the water may be calculated, as shown in the following equation:

$$pH_s = A + B - \log Ca^{2+} - \log M$$

where A (°F or °C) and B (dissolved solids) are obtained from Table 3.2. The calcium (Ca^{2+}) and methyl orange alkalinity (M) (as calcium carbonate) are obtained from the water analysis in question. The difference between the actual pH of the water supply and pH_s [pH of water saturated with calcium carbonate ($CaCO_3$)] is the Saturation Index.

Assuming that a water contains a Ca^{2+} (as $CaCO_3$) concentration of 200 mg/L, an M alkalinity (as $CaCO_3$) concentration of 60 mg/L, and a total dissolved solids concentration of 650 mg/L and that the pH mea-

TABLE 3.2 Values of A and B Necessary for Calculation of the Langelier Saturation Index

Temperature						Dissolved solids			
°F	°C	A	°F	°C	A	mg/L	B	mg/L	B
32	0	2.60	104	40	1.71	0	9.70	500	9.86
41	5	2.47	113	45	1.63	25	9.73	600	9.87
50	10	2.34	122	50	1.55	50	9.76	700	9.88
59	15	2.21	131	55	1.48	75	9.78	800	9.89
68	20	2.10	140	60	1.40	100	9.80	900	9.90
77	25	1.98	149	65	1.34	125	9.81	1000	9.91
86	30	1.88	158	70	1.27	175	9.82	1100	9.92
95	35	1.79	167	75	1.21	225	9.83	1200	9.93
			176	80	1.17	300	9.84	1300	9.94
						400	9.85	1400	9.95

sured at the site was 9.0 at a temperature of 16°C (61°F) the pH_s and the Saturation Index are calculated as follows:

$$\log 200 = 2.30; \log 60 = 1.78$$

$$pH_s = 2.19 + 9.88 - 2.30 - 1.78 = 7.99 \text{ (or, rounding off, 8.0)}$$

$$\text{Langelier Saturation Index (LSI)} = 9.0 - 8.0 = +1.0$$

The positive value indicates that the water is supersaturated with respect to $CaCO_3$; if the Saturation Index result had been zero, then the water would be stable, not tending to form scale or dissolve $CaCO_3$; if the result had been negative, then the tendency to dissolve $CaCO_3$ and a possible corrosive tendency would have been indicated.

It should be recognized that the above simplified method of calculation does not take into account the possible presence of ion pairs (such as $CaHCO_3^+$) and their effect on the calcium concentration. The more complicated method of calculation is usually not required.[17-19]

In the presence of scale inhibitors, such as the polyphosphates, phosphonates, and acrylates, which have a pronounced effect on the solubility of calcium carbonate, the Saturation Index values are not informative with respect to scaling and corrosive tendencies. As expected, the calcium carbonate Saturation Index or equivalents will not yield information as to the degree of scale formation in the form of calcium sulfate, calcium phosphate, magnesium hydroxide, or silica. Scaling tendencies for these compounds must be determined from their own solubility relationships. Also, the calcium carbonate Saturation Index cannot be expected to yield complete information on corrosiveness when other more pertinent factors, such as dissolved oxygen, chloride, sulfate, buffer capacity, and velocity, are not considered.

Singley[19] has reported that the Langelier Saturation Index (LSI) is not a reliable indicator of the corrosive tendencies of a water and that an equation including the chloride, sulfate, alkalinity, dissolved oxygen, buffer capacity, calcium, LSI, and length of time of exposure provides a good indication. This equation for determining corrosive tendencies is as follows (in computer language, ^ means raised to power, and * stands for multiplication):

$$JS = (CL^\wedge.509) * (SO^\wedge.0249) * (TA^\wedge.423) * (O_2^\wedge.78)$$

$$RP = ((CA \text{ (as } CaCO_3) * 0.4)^\wedge.676) * (BI^\wedge.0304) * ((10^\wedge LSI)^\wedge.107) * 9.4675$$

where CL = chloride
SO = sulfate

TA = total alkalinity (as $CaCO_3$)
O_2 = dissolved oxygen
CA = calcium (as $CaCO_3$)
BI = buffer intensity
LSI = Langelier Saturation Index

Then:

$$\text{Singley Corrosion Rate (SG)} = JS/RP$$

This can be easily calculated using the computer programs shown in App. D.

By using a computer to calculate the various calcium carbonate indexes obtained by the computer programs in App. D, combining information from various corrosion studies in the chemical literature, and using information from his or her own practical experience, the water treatment consultant can come forth with reasonably good judgment as to the expected corrosive and scaling tendencies of a water supply.

$CaSO_4^0$, $CaCO_3^0$, and $MgCO_3^0$ (ion pair complexes) are soluble but not ionized and therefore not effective in the chemical equilibriums associated with calcium carbonate solubility. Their presence is largely responsible for the need to add +0.4 to +1.0 to the calculated LSI pH_s value for waters of low hardness and alkalinity and waters of high magnesium and sulfate content in order to provide a stable treated water. This difference between the required pH and the pH of saturation (pH_s) varies with different waters; however, there are several computer programs available that may be referred to for this calculation. In typical Illinois waters, the complexes cause apparent shifts in pH_s and Saturation Index of from 0.02 to 0.3 unit.

Ryznar Index

In addition to the Langelier Saturation Index (LSI), there is the Ryznar Index (RI),[20] which is calculated as follows:

$$RI = (2 \times pH_s) - pH$$

The Ryznar Index yields values below 6.0 if scaling tendency is indicated, and a tendency to dissolve $CaCO_3$ is indicated at values above 6.0. While this index has an empirical origin, it provides a reasonably good quantitative measure of the amount of scaling to be expected, and it is still being used to estimate expected scale formation even in the presence of inhibitors such as the polyphosphates, phosphonates, and acrylates in cooling waters. This the author can verify, since he did the laboratory work when this index was developed. Figure 3.2 shows the good agreement of the Ryznar Index with observed scaling conditions.

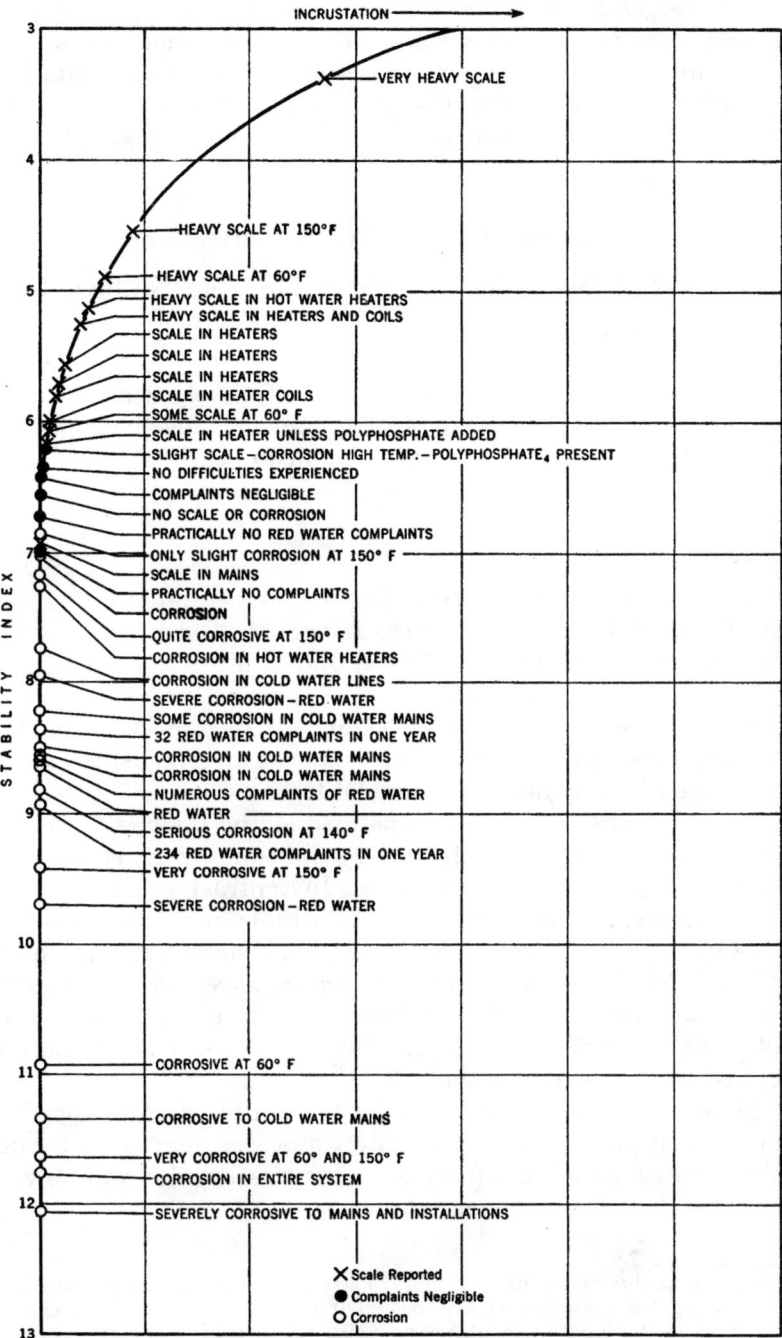

Figure 3.2 Ryznar Index vs. scale results. (*Nalco Chemical Co., Chicago, Ill.*)

It should be recognized that the Langelier Saturation Index (LSI) value reveals only a tendency to form scale and does not provide a quantitative evaluation of the amount of scale expected to be formed. Data from the original paper introducing the Ryznar Index[20] disclosed that the amount of scaling could be calculated from the following equations:*

Within RI values of 2–7:

$$\text{milligrams scale} = 44700/(2pH_s - pH)^{4.73}$$

Within RI values of 2–4 in presence of polyphosphate ($Na_9P_7O_{22}$):

$$\text{milligrams scale} = 631000/(2pH_s - pH)^{8.65}$$

This water-testing technique has been used effectively to demonstrate that waters of high hardness can be treated with a few parts per million (mg/L) of polyphosphate to minimize calcium carbonate scaling at moderate temperatures.

Other indexes

Others have come forth with the Driving Force Index (DFI), Momentary Excess (ME), Aggressiveness Index (AI), Calcium Carbonate Precipitation Potential (CCPP), Singley Index, etc., to characterize a water's corrosiveness or scaling tendency. In general, practically all are based on the solubility of calcium carbonate and its protective qualities in decreasing corrosion. There are redundancies involved, since waters with the same LSI may have the same DFI.

The *Driving Force Index (DFI)* is defined as the product of the calcium and carbonate activities divided by the equilibrium constant K_s for the solubility of calcium carbonate. In contrast to pH_s and the Saturation Index, the DFI is not a logarithmic quantity; therefore, large changes in the DFI are associated with small changes in the Saturation Index. In the case of the *Momentary Excess (ME)*, a positive value indicates oversaturation, zero indicates a saturated water, and a negative value indicates an undersaturated water with respect to $CaCO_3$, just the same as with the LSI.[20–22]

The *Aggressiveness Index (AI)* (similar to the LSI) was developed for asbestos-cement pipe to show whether a water was aggressive to such pipe. Waters with an AI less than 10 are aggressive, and waters with an AI greater than 12 are nonaggressive.

---

*The results were obtained using tared glass coils immersed in a heated glycerine bath through which 2 gal of treated water were passed at a 1-gal/h flow rate and at an effluent temperature of 200°F (93°C). While not a precise measurement of scaling tendency, the calculation provides a semiquantitative method of measuring scaling tendency.

The *Calcium Carbonate Precipitation Potential (CCPP)*[17,19] is an informative index because its value actually provides a measure of the *quantity* of calcium carbonate ($CaCO_3$) that will be precipitated or dissolved in *a water of known analysis. However, the calculation of this index is rather complex, requiring a computer program* as shown in App. D.

Though based on laboratory investigations, the *Singley Index*[19] is more meaningful in predicting the corrosion rate because the calculation of this index includes the chloride, sulfate, alkalinity, dissolved oxygen, calcium, buffer capacity, length of exposure time, and $CaCO_3$ Saturation Index (LSI).

Riddick's Index is another index that apparently provides dependable information on the degree of corrosiveness of low hardness waters in the northeastern United States. It is calculated as follows:

$$\text{Index} = 75/M * [CO_2 + 0.5 * (H - M) + Cl^-]$$
$$+ 2 * (NO_3^-) * (10/SiO_2) * (DO + 2/DO_{sat})$$

in which H (hardness), M (alkalinity), and CO_2 are provided as $CaCO_3$, chloride (Cl^-), nitrate (NO_3^-), silica (SiO_2), dissolved oxygen (DO), and saturated dissolved oxygen (DO_{sat}) at the temperature of measurement.

Index = <5, very noncorrosive; = 6–25, noncorrosive

Index = 25–50, mildly corrosive; = 51–75, moderately corrosive

Index = 76–100, corrosive; = >100, very corrosive

Buffer intensity,[23] which is a quantitative measure of the acid or alkali required to produce one unit change in pH in a subject water, is also calculated to attain a more complete picture of the scaling or corrosive tendency of a water. Its meaning is appreciated when one realizes that during electrochemical corrosion reactions, localized anode and cathode areas are produced on the surface of the metal. Acid and alkaline conditions are produced at these areas, and if the buffer intensity of a water is high enough, it may counteract these localized accumulations and decrease the overall electrochemical reaction and rate of corrosion. The calculation of buffer intensity is best done by using the computer programs in App. D.

The *Larson Index*[22] is calculated from the relative ratio of the total of chloride and sulfate ions to the total alkalinity of a water as follows:

$$(\text{mequiv/L } Cl^- + \text{mequiv/L } SO_4^{2-})/\text{mequiv/L M alkalinity}$$

where mequiv/L is milliequivalents per million. The lower the value, the less corrosive the water; a value less than 0.2 is considered to mean that a water is relatively noncorrosive.

Low alkalinity waters (<50 mg/L) are usually associated with high corrosion rates for the following reasons:

1. Low pH, low carbonate (CO_3), or other anions prevent protective scale formation.
2. Low buffer intensity associated with low alkalinity waters causes localized pH differences and increased corrosion.
3. Low conductivity tends to localize corrosion cells and increase pitting tendencies.
4. Low pH increases the OH^- gradient near cathodes and increases diffusive transport.

At the Conference of the National Association of Corrosion Engineers (NACE) in St. Louis in 1988, the author presented a paper[24] that included the same computer programs as in App. D for calculating most of the above indexes and for determining the quantity of chemicals required to adjust indexes to a desired value. For example, in the case of an eastern U.S. major municipality water of low hardness and low alkalinity, these computer programs can be useful in calculating the amount of lime and carbon dioxide required to make the water less corrosive. It would be desirable to increase the calcium and alkalinity to at least 40 ppm (mg/L) to make the water less corrosive and more stable, with a reasonable pH.

The water chemist who has the values of at least six different indexes of a water is in a much better position to predict water corrosiveness or scaling. It has been learned that some of the indexes are more informative with respect to waters of different qualities. For example, it has been observed that the Larson Index[22] provides the best evaluation of corrosive tendencies of a water like that in Portland, Oregon, which is a very low hardness and solids water derived from mountain snows.

Metallic Contamination of Drinking Water

The Safe Drinking Water Act (SDWA) prepared by the U.S. Congress makes the water utility responsible for water quality at the customer's tap. The water purveyor is therefore responsible for properly treating the water treatment plant effluent so that corrosion does not occur in the small diameter (high surface/volume ratio) of galvanized, steel, copper, and soldered copper in building water systems. Such corrosion can cause dissolution of metals in the piping system to the extent of exceeding the MCL (maximum contaminant level) specified by the U.S.

Environmental Protection Agency. Proper treatment is also important because corrosion products may provide harboring areas where the growth of bacteria may be promoted, the effectiveness of chlorination may be limited, and poor-quality drinking water may result.

In 1986, an amendment to the SDWA was passed by the U.S. Congress banning the installation of lead in public water systems and in residences connected to these supplies. This amendment to the SDWA forbids the use of pipe, solder, or flux that is not lead-free for the installation or repair of public water supplies or plumbing systems providing water for human consumption. *Lead-free* means that solder and flux shall not contain more than 0.2 percent lead and that pipes and fittings shall not contain more than 8 percent lead.

In addition, the water purveyor must:

1. Ensure that water entering the water distribution system has an MCL of 5 µg/L (ppb) or less lead and 1.3 mg/L (ppm) copper
2. Implement a public education program with regard to lead contamination
3. Foster a corrosion control program to reduce the corrosiveness of the water supply[1,2]

This three-pronged approach to the Lead and Copper Rule involves treatment, public education, and lead service line replacement. The water purveyor must conduct materials surveys of the distribution system and interior plumbing conditions within its service area and must select appropriate sample sites and representative locations for sampling of the distribution system water to check the quality. Studies must be performed to determine optimal treatment and to implement the following action levels (ALs):

Lead AL: ≤0.015 mg/L in at least 90% of the samples

Copper AL: ≤1.3 mg/L in at least 90% of the samples

If systems still exceed the ALs in follow-up monitoring, the purveyor may be required to exert further efforts; for example:

1. Continue the public education program.
2. Begin a lead service line replacement program.
3. Continue a monitoring program until ALs are met.

While the above rule pertains mainly to the water purveyor, the building manager should consider sampling and the analysis of first-draw samples (such samples can likely be analyzed by the local water purveyor) in order to be assured that metallic contaminants are not in ex-

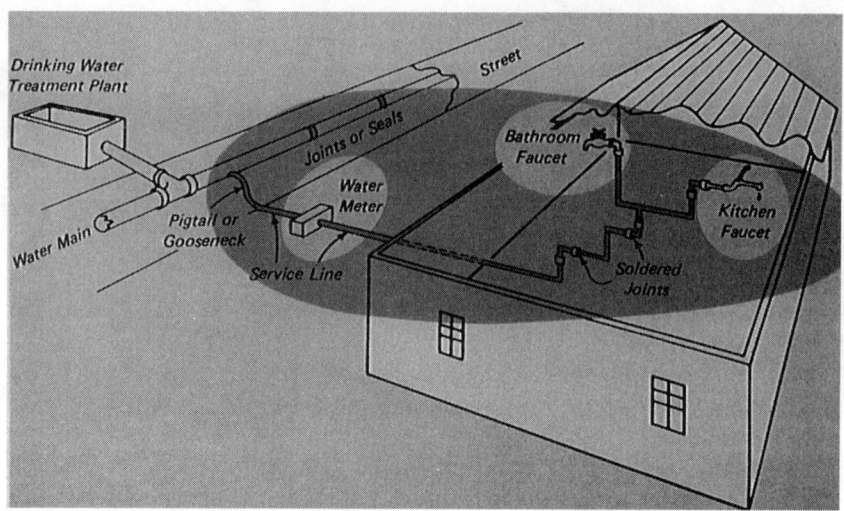

Figure 3.3 Sources of lead in public water system and homeowner plumbing. (*U.S. Environmental Protection Agency.*)

cess. Figure 3.3 shows the possible sources of lead (pigtail, soldered joints, and faucets) in the public water system and homeowner plumbing.

Building managers should have concern for the quality of the consumers' drinking water and should desire to comply with the National Primary Drinking Water Regulations for Lead and Copper. In this regard, they must decide what strategy to pursue to eliminate a lead contamination problem. Since the following items may affect the decision, attention must be given to each one:

1. The age of the building
2. The piping metal and solder used in the building
3. The velocity, residence time, temperature, and water analysis (pH, hardness, alkalinity, dissolved oxygen)

Evidence indicates that tin-lead solder in a copper piping system may likely be a more serious source of lead contamination than the lead pigtail often used in connecting building services to the water distribution main. High velocities (above 1 m/s) may cause increased corrosion of both copper and lead-bearing surfaces. Long residence, or stagnation, time (4 to 24 h or longer) has been observed to cause higher lead or copper contamination levels in the drinking water. Also, at higher water temperatures, somewhat higher lead contamination can be expected.

With regard to the analysis of the water from the treatment plant, the following statements can generally be made:

1. Waters with a pH below 7.5 will generally be corrosive, and waters with a pH above 7.5 will be less corrosive.
2. Waters of hardness of 120 ppm or more will generally be less corrosive than waters of lower hardness.
3. Waters of alkalinity of 60 ppm or below will generally be less corrosive to lead if the pH is 7.5 or above.
4. Increased dissolved oxygen (4 mg/L and above) will generally be expected to cause increased corrosion of lead.

The major source of lead in drinking water is not the source water but the corrosive action of the water on materials in the building or resident's distribution system. High lead contamination in the drinking water of buildings that have newly installed tin-lead–soldered copper piping is most likely to occur during the first 2 years. After this period, the lead may gradually decrease, and after 5 years, it may become even less of a problem.

Development of a protective film apparently causes a reduction in the solubility of the lead in tin-lead–soldered joints and lead-containing valves and tanks. Before 1986, the tin-lead solder generally used for joining copper tubing was known as 50:50 tin-lead solder (50 percent tin, 50 percent lead). Since lead is anodic to copper, it is not surprising that the lead in solder, being a small anode and in contact with a large copper cathode, is dissolved in drinking water.

Substitution of 95:5 tin-antimony and 95:5 tin-silver solders is now being found satisfactory in joining copper piping. Although these solders are more difficult to apply because of their higher melting temperatures, their corrosion resistance is superior.

Health problems related to metallic contamination

Why are we so concerned about the lead levels being found in drinking water? Ingestion of lead is reported to cause severe damage to the brain, kidneys, nervous system, and red blood cells, even at levels below 0.050 mg/L (the 1989 MCL), and so lowering the limit to 0.015 µg/L or below is definitely justified. Lead has been blamed for elevated blood pressure, strokes, heart attacks, and hypertension in adults for some time. High lead levels may be particularly bad for pregnant women, causing complications during pregnancy and possibly damaging the fetus. Also, there is evidence that high lead levels cause retardation in children.[2]

Sources of lead may be the air, food, dust, and drinking water. Lead in the air is derived mainly from the combustion of leaded gasoline or from industrial emissions. A health hazard for children has been the ingestion of lead paint chips found in older homes. The U.S. Environmental Protection Agency estimates that the average two-year-old child may get about 20 percent of his or her total lead exposure from drinking water.

The lead found in our drinking water is derived mainly from the corrosion of goosenecks, lead service lines, lead plumbing, lead-lined iron piping, 50:50 tin-lead solder, and brass faucets. Lead piping and goosenecks connecting home services to the municipal distribution system are prevalent in the eastern United States, but tin-lead–soldered copper pipe is prevalent throughout the entire country. The levels of lead found in the drinking water are appreciably affected by the age of the pipe, type of material, quality of work, size of pipe, water quality, size of the water sample, standing time, and type of sample (standing or running sample).

Relevance of sampling methods

Water sampling techniques are necessary because the USEPA has based its limits on first-draw samples from the consumer's tap after the water has been allowed to stand 8 h overnight with no flow. In general, running, or flowing, samples (800 to 1,000 mL) taken later, after there has been significant water usage, usually are found to be appreciably below the MCL. However, this may not apply to a very corrosive water [low pH of 4 to 5; high chlorides and sulfates (500+ mg/L)] or a water that does not meet the alkalinity-pH concept prescribed previously.

Water faucets themselves can also contribute appreciable lead to drinking water, as the brasses involved in their manufacture contain as much as 8 percent lead. Lead is required in brass so that the valve metal can be properly machined. Since the flow through valves is turbulent, it should not be unexpected that lead, being a component of the exposed metal, will dissolve in the water passing through the valve. Since water is exposed in a standing, or stagnant, condition in faucets between uses, the first water drawn may be expected to be high in lead. Therefore, advice is given to allow water to flow from the faucets until it approaches the cold water temperatures expected for water in the city mains; then water can be drawn for drinking or food preparation. Flushing water piping before use is to be encouraged, as allowing water to stand in piping for long periods increases the solubility of the metals exposed to the water. All-stainless-steel valves are now available for replacement of brass valves.

The public has been warned for some time not to drink the water first coming from a tap,[10] as the flavor is often impaired. Coffee prepared from this first-draw sample may have an off-flavor.

Besides tin-lead solder and brass valves, another serious source of lead is the lead-lined holding tanks in drinking-water coolers. Manufacturers that provide lead-lined storage tanks as part of their water coolers have been requested to discontinue this practice; they have arranged for replacement with lead-free equipment.

It is interesting to note that New Zealand and Australia recognized the potential lead corrosion and drinking-water-contamination problem from tin-lead–soldered copper tubing years ago and passed legislation preventing the use of tin-lead solder in the construction of potable water systems.

Methods of reducing metallic contamination

The solution to reducing the lead levels in building systems when they exceed the specified MCL (maximum contaminant level) is to arrange first for sampling and analysis of the influent water to learn whether the metallic contamination is from a source outside the building. If the influent water is free of metallic contamination but building water samples are above the MCL for lead, then consideration should be given to applying chemical treatment for correcting the problem. The addition of alkaline chemicals to raise the pH and reduce the corrosiveness of the metals in the system may be necessary. A consultant's advice is recommended to learn which chemical and/or corrosion inhibitor is appropriate for application. Increasing the pH of the supply to near 8.0 or adjusting the Langelier Index to a positive value (see the section on indexes in this chapter) has proved to be one of the best methods for reducing the corrosiveness of many supplies.

High chlorine (above 1 ppm), which some municipalities apply to be assured of adequate chlorination, has been observed to cause an increase in the lead concentration. The amount of chlorine may likely be reduced without risk of inadequate chlorination.

Application of orthophosphate to 0.5 to 1.0 mg/L as phosphate (PO_4) either in sodium or in zinc form has reduced lead contamination. The advantage of applying zinc orthophosphate rather than sodium orthophosphate is that the zinc provides corrosion inhibition for the copper and steel in the system. Dosages of zinc should probably be limited to about 0.5 to 2.0 ppm, as some health experts in other countries believe that zinc may be injurious to the public health, although the USEPA MCL is 5 ppm.

As excess flux and flux residue in new systems may have been contributing to the metallic contamination problem, it is now recognized that these systems should be properly cleaned by circulating hot water [140 to 150°F (60 to 66°C)] through them.[25] Hot water of this temperature is a much better solvent of flux residue than cold water.

In one case with which the author was involved, the pH was high (above 8.6). Since this is above the limit for the most effective zinc phosphate treatment for controlling lead contamination, it was necessary to apply carbon dioxide to reduce the pH to near 8.0. In another case, the pH of the incoming water for a high-rise building was only 7.0, with a low alkalinity of 10 ppm, and so it was considered necessary (*a*) to decrease the corrosive tendencies of this water by raising the pH to 7.5+ by adding soda ash and (*b*) to add zinc orthophosphate to provide 0.5 to 1.0 ppm orthophosphate (as PO_4) to effectively reduce the lead contamination.

Copper, not being very toxic, does not present a real contamination problem. However, the MCL for copper may eventually be lowered to 1.0 ppm (mg/L). Copper is relatively corrosion-resistant but is corroded by acidic waters [low pH (4.0 to 7.0)] and will be found in 1- to 5-mg/L quantities in copper or brass systems in standing samples after overnight nonusage. Residence time, water analysis variables (such as carbon dioxide, ammonia, and chlorine), and velocity may be expected to influence the quantities found. In one instance, it was observed that copper tested 3 to 5 ppm (mg/L) in a red brass piping system in a high-rise building; however, this building had never experienced a leak or corrosion in its 60-year life. A more complete discussion of the corrosion of copper is in Chap. 2.

Metallic contamination other than by lead and copper

Cadmium derived from impurities in galvanized pipe has generally not exceeded the MCL of 10 ppb in drinking water.

Drinking water exceeding the MCL of 5 ppm for zinc is sometimes found where the water is exposed to galvanized piping. A discussion of the corrosion resistance of galvanized-steel piping is included in Chap. 2.

Drinking water containing iron even above the MCL of 0.3 ppm is not considered to be toxic. Its main disadvantages are the off-flavor, the discoloration of the water, and the brown stain observed in lavatories and on laundered clothes. Manganese also shares these disadvantages.

Although there have been reports linking aluminum in drinking water to Alzheimer's disease, leading medical experts[26] and the U.S. Food and Drug Administration and Environmental Protection Agency (EPA) all concur that aluminum is not a risk factor. Exposure to aluminum in drinking water is very small (1 percent) compared with its general prevalence as the third most abundant element in the earth's crust (after oxygen and silicon).

In the past, the literature indicated that sodium or salt in drinking water caused hypertension; however, it is now known[27] that less than 10 percent of the sodium we consume in food comes from drinking water. The sodium in zeolite-softened water had been of concern because the hardness content of water is converted to sodium salts in the ion exchange process. As a result of this more comprehensive understanding, the U.S. EPA, as well as several states and organizations, has now removed sodium from its list of regulated drinking-water contaminants.

There has been concern over nitrates in drinking water[28] since 1945, when a blood disorder called methemoglobinemia was observed to cause "blue baby syndrome." Changes in nitrate in the digestive system of infants up to six months old inhibit oxygen transport to the blood. Nitrate contamination is prevalent in shallow wells that receive runoff from fertilized agricultural fields, animal pens, and septic tanks. Nitrates may accumulate in subsoils and leach into aquifers. An MCL of 10 mg/L as N (nitrogen) is specified. Effective removal can be provided by distillation, electrodialysis, and reverse osmosis. Recently, an anion resin* that removes nitrate in preference to other anions, such as sulfate, has been used effectively.[29] In areas where the nitrate is prevalent and may vary in concentration, tests should be conducted at least every 6 months to determine whether the present level is below the specified MCL.

References

1. M. M. Frey, L. L. Harms, and H. A. Susia, "Complying with the Lead and Copper Rule," *WATER/Engineering & Management,* August 1991, p. 22, and F. W. Pontius, "A Current Look at the Federal Drinking Water Regulations," *Journal of the American Water Works Association,* March 1992, p. 36.
2. *Lead Control Strategies,* American Water Works Association Research Foundation, Denver, 1990.
3. "Corrosion and Corrosion Control in Drinking Water Systems," *Proceedings from a Corrosion Workshop and Seminar in Oslo,* National Association of Corrosion Engineers, Houston, Mar. 19–21, 1990.
4. *Equipment Handbook,* chap. 38: "Drinking Water Coolers and Central Systems," American Society of Heating, Refrigeration and Air-Conditioning Engineers, Atlanta, 1988.
5. *Equipment Handbook,* chap. 40, "Automatic Ice Makers," American Society of Heating, Refrigeration and Air-Conditioning Engineers, Atlanta, 1988.
6. B. J. Brazos, J. T. O'Connor, and S. Abcouwer, "Kinetics of Chlorine Depletion and Microbial Growth in Household Plumbing Systems," *Proceedings of the American Water Works Association Water Quality Technology Conference,* Houston, 1985.
7. J. F. Wilkes, "Toilet Tank Travail—Accelerated Failures of Rubber Components in Florida Potable Water Supplies," Preprinted Extended Abstract, American Chemical Society, Division of Environmental Chemistry, Miami Beach, September 1989.

---

*Ionac SR-6 resin, manufactured by Sybron Chemicals, Inc., Birmingham, N.J., is reported to remove nitrate and to pass sulfate.

8. T. Troxell, "Bottled Water: Are Current Regulations Sufficient?" *U.S. Water News,* December 1990, p. 7.
 9. H. Allen, C. Haas, and M. A. Henderson, "Analysis of Bottled Waters," *Chemtech,* December 1991, p. 738.
10. C. Ingram, *The Drinking Water Book,* Ten Speed Press, Berkeley, 1991.
11. *Standards Number 60 and 61, Drinking Water Chemicals and System Components—Health Effects,* National Sanitation Foundation, Ann Arbor, 1988.
12. R. W. Lane, T. E. Larson, C. H. Neff, and S. W. Schilsky, "Silicate Treatment Inhibits Corrosion of Galvanized Steel and Copper Alloys," *Materials Performance,* April 1973, pp. 32–37.
13. R. W. Lane, T. E. Larson, and S. W. Schilsky, "The Effect of pH on the Silicate Treatment of Hot Water in Galvanized Piping," Illinois State Water Survey Reprint No. 367, Champaign, 1977.
14. *Test Methods for Corrosivity of Water in Absence of Heat Transfer (Weight Loss Methods),* D2688-90, *1991 Annual Book of ASTM Standards,* vol. 11.01, American Society for Testing and Materials, Philadelphia, p. 218; also *Test Methods for Corrosivity of Water in Absence of Heat Transfer (Electrical Methods),* D2776-79, p. 207.
15. *Prevention & Control of Water-Caused Problems in Building Potable Water Systems,* TPC Publication 7, National Association of Corrosion Engineers, Houston, 1980.
16. W. F. Langelier, "The Analytical Control of Anti-Corrosion Water Treatment," *Journal of the American Water Works Association,* vol. 28, October 1936, p. 1500.
17. J. R. Rossum and D. T. Merrill, "An Evaluation of the Calcium Carbonate Saturation Indexes," *Journal of the American Water Works Association,* vol. 74, no. 2, 1983, p. 95.
18. Joint Task Group on Calcium Carbonate Saturation, "Suggested Methods for Calculating and Interpreting Calcium Carbonate Saturation Indexes," *Journal of the American Water Works Association,* July 1990, p. 71.
19. R. A. Pisigan, Jr., and J. E. Singley, "Evaluation of Water Corrosivity Using the Langelier Index and Relative Corrosion Rate Models," *Corrosion/84,* Paper 149, National Association of Corrosion Engineers, New Orleans, 1984.
20. J. W. Ryznar, "A New Index for Determining Amount of Calcium Carbonate Scale Formed by a Water," *Journal of the American Water Works Association,* vol. 36, April 1944, p. 472.
21. R. W. Lane and C. H. Neff, "Indices for Estimating Corrosive and Scale Tendencies of Water," Tenth Annual Electric Utility Workshop, Urbana, Ill., 1990.
22. T. E. Larson, *Corrosion by Domestic Waters,* Illinois State Water Survey Bulletin 59, Urbana, 1975.
23. J. M. Montgomery, Consulting Engineers, Inc., *Water Treatment Principles and Design,* John Wiley & Sons, New York, 1985, p. 423.
24. R. W. Lane, "Using Computer Programs to Control Corrosion and Scale in Water Systems," *Materials Performance,* vol. 27, no. 9, September 1988, p. 26.
25. V. L. Van Blaricum, O. S. Marshall, R. H. Knoll, and V. F. Hock, "A Rehabilitation Technique for Soldering Flux-Induced Corrosion in Copper Potable Water Piping Systems," *Corrosion/91,* Paper 339, Cincinnati, March 1991.
26. R. D. Harriger and J. C. Steelhammer, *Alzheimer's Disease; Aluminum and Drinking Water Treatment; A Review,* General Chemical Corp., Syracuse, N.Y.
27. R. L. Hanneman, "Evidence Challenges Sodium-Health Connection," *Water Technology,* October 1989, p. 30.
28. G. S. Ellis, "Removing Nitrates from Drinking Waters," *Water Technology,* August 1990, p. 36.
29. Gregg and Brown, "Nitrate Removal from Municipal Well Water," 34th International Water Conference, Pittsburgh, October 1973, p. 179.

Chapter

4

Steam and Hot Water Heating Systems

Buildings are usually heated by hot water or steam circulated in pipes from a central boiler room to the areas requiring heat. In many building systems, steam is produced at 5 to 400 psig pressure in boilers and then is piped to a nearby heat exchanger to heat water to the desired temperature for heat distribution systems, for domestic uses, or for many other purposes. Such heat transfer systems are described as hydronic systems.[1] Hot water has advantages over steam in conveying heat to needed areas:

1. More heat (Btus) can be carried in the same-sized piping.
2. Steam traps are not required.
3. Water treatment and control are much simpler.

An advantage of steam is low density for transporting heat to tall buildings. Also, steam serves as a by-product of electrical production and is a somewhat more reliable heat source.

The most common boilers are fossil-fuel-fired. These boilers are normally heated by burning oil or gas or coal, and they may have fluidized-bed combustors for desulfurizing flue gases. Presently, cogeneration is commonly being installed by building owners, since substantial savings can result from simultaneously producing electricity and providing building thermal requirements from the same energy source. Much of the heat generated by a prime mover producing electricity may be used to energize absorption chillers and to provide low-pressure steam or hot water for space heating, for heating domestic hot water, or for other process heat requirements. An example is a building heating system serving as a condenser for the low-pressure steam released from an operating noncondensing steam turbine that generates electricity.

Regulations

The American Society of Mechanical Engineers (ASME) has formulated a Boiler and Pressure Vessel Code,[2] which serves as a guide for the construction and operation of steam and hot water boilers. In addition, maintenance and repair facilities are inspected by the National Board of Boiler and Pressure Vessel Inspectors, and these facilities are certified by the National Board if they meet the board's regulations. The ASME code has been adopted by state, county, and city governments to assure that pressure vessels are operated safely. Regular and periodic internal inspections are required each year to ensure that the boiler is safe to operate and that proper preventive maintenance is being administered; otherwise, the permit to operate the equipment may be terminated.[2]

General Steam Power Plant Operations[3-7]

Feedwater fed to a steam boiler normally consists of return steam condensate and makeup water required to replace losses of condensed steam from the system. Building (closed) steam or hot water systems should require a desired minimum addition of makeup water. This can be accomplished if there is adequate maintenance of distribution piping and mechanical equipment, as well as application of proper water treatment and conscientious water treatment control. The use of minimum makeup water ensures that water-exposed equipment can be more easily kept free of problems that could result from corrosion and scale formation. Industrial boiler water systems, however, may require 25 to 90 percent of the steam system's capacity for process steam use. Since this steam is not returned to the system as condensate, these systems usually require more complete external water treatment equipment, such as dealkalizers, water softeners, and specific internal water treatment chemicals to ensure that high-quality makeup water is provided for these systems.

In most building systems, process steam demands are expected to be minimal, though steam may be required for humidification, sterilizers, or laundry or kitchen uses. If these uses are appreciable, design of the water treatment system will require the installation of the more extensive and complete treatment equipment. In the normal building water system, steam boiler blowdown requirements should amount to less than 5 percent, and if there is a deaerator in the system, a few additional tenths of 1 percent loss through the deaerator vent may be expected. There may also be process steam losses as a result of laundry, kitchen, and humidification operations and miscellaneous system losses, such as from leaks and tank overflow. If there is no process steam usage, the total makeup water requirements should be less than 10 percent. It is important to keep makeup water usage low because added water solids and gases derived from increased makeup water usage can cause an increase in deposition and corrosion in the boiler and accompanying water systems.

Conscientious water treatment control and monitoring are essential, even if the water source is of high quality and proper water treatment equipment has been installed. Without proper water treatment, energy loss equivalents can be substantial, as shown in Table 4.1. Besides the loss in efficiency, boiler tube corrosion, perforation, and ruptures may result from the development of appreciable scale and corrosion. Such failures are costly and often cause serious interruptions of needed heat and power.

Proper water treatment and control of the whole boiler system must be considered as part of the care and maintenance required for this power-producing equipment. *The whole system* means the makeup water system, the feedlines, the economizer, the boiler, the turbine, the turbine condenser, the return condensate system, the closed heating and cooling systems, and the open cooling system. Each part of the whole system is exposed to a significantly different environment with respect to the pH level, alkalinity, and conductivity, as well as temperature and pressure. In the initial design of the system, plans should include conducting proper control testing at numerous specific locations in order to attain optimum water treatment and operating efficiency. Condensate probably has the lowest conductivity and pH; next is the feedwater; and highest in conductivity and pH is the boiler water. There is a definite optimum pH, alkalinity, inhibitor concentration, and conductivity at which each part of the boiler system can best be kept corrosion-resistant. Therefore, the overall water treatment plan

TABLE 4.1 Energy Loss Equivalent Resulting from Scale Formed on Waterside of Boiler Heat Transfer Surface[7]

Waterside scale thickness, in	Energy loss equivalent, %
1/16	15
1/8	20
1/4	39

SOURCE: *Plant Engineering,* May 24, 1990.

must be designed to satisfy these individual requirements as well as the system as a whole.

Boiler Water Treatment

Benefits to be expected from a well-designed boiler water treatment program are as follows:

1. Reduced fuel consumption
2. Reduced maintenance of the boiler and distribution system
3. Fewer outages and disruptions of operations
4. Increased safety
5. Reduced water and chemical costs
6. Extended equipment life

The scientific literature yields information[8] on water treatment requirements for the many uses of water, as shown in Table 4.2. However the American Society of Mechanical Engineers (ASME) provides much more specific advice on operating practices for the control of feedwater and boiler water quality in modern industrial boilers[9] (see Table 4.3).

Figure 4.1 is an informative diagram of the treatment scheme for applying internal water treatment chemicals to the preboiler, boiler, and condensate systems to attain the optimum and required corrosion and scale control for efficient operation.

The objectives of boiler water treatment are as follows:

1. To control scale formation and dispersion and to control sludge accumulations
2. To inhibit corrosion in boilers and connecting water systems
3. To inhibit stress corrosion cracking
4. To maintain high steam purity by control of boiler water carryover.

Each of the above objectives will now be discussed in detail.[10–12]

Control of scale formation

The solubility of water constituents must be given consideration, since accumulated water hardness resulting from evaporation may be precipitated on heating and may deposit as scale or sludge on heat transfer surfaces. Water treatment is necessary to prevent scale and sludge from causing lowered heat transfer efficiency, metal deterioration, and

TABLE 4.2 Water Quality Requirements (mg/L) Are Most Stringent for High-Pressure Boiler Feedwater; Requirements for Many Other Uses Can Be Achieved by a Well-Run Secondary Treatment Plant

Characteristics	Boiler feedwater, psig				Cooling water			
					Once-through		Makeup for recirculation	
	0–150	150–700	700–1500	1500–5000	Fresh	Brackish	Fresh	Brackish
Silica (SiO$_2$)	30	10	1.0	0.01	50	25	50	25
Aluminum (Al)	5	0.1	0.01	0.01	—	—	0.1	—
Iron (Fe)	1	0.3	0.05	0.01	—	—	0.5	—
Manganese (Mn)	0.3	0.01	0.01	—	—	—	0.5	—
Copper (Cu)	0.5	0.05	0.05	0.01	—	—	—	—
Calcium (Ca)	—	0	0	*	200	420	50	420
Magnesium (Mg)	—	0	0	*	—	—	—	—
Sodium and potassium	—	—	—	—	—	—	—	—
Ammonia (NH$_3$)	0.1	0.1	0.1	0.7	—	—	—	—
Bicarbonate (HCO$_3$)	170	120	50	*	600	—	25	—
Sulfate (SO$_4$)	—	—	—	—	680	2,700	200	2,700
Chloride (Cl)	—	—	—	—	600	—	500	—
Fluoride (F)	—	—	—	—	600	19,000	500	19,000
Nitrate (NO$_3$)	—	—	—	—	—	—	—	—
Phosphate (PO$_4$)	—	—	—	—	—	—	—	—
Dissolved solids	700	500	200	0.5	1,000	35,000	500	35,000
Suspended solids	10	5	0	0	5,000	2,500	100	100
Hardness (CaCO$_3$)	20	1.0	0.1	0.07	850	6,250	130	6,250
Alkalinity (CaCO$_3$)	140	100	40	0	500	115	20	115
Acidity (CaCO$_3$)	—	—	—	—	—	—	—	—
pH (units)	8.0–10.0	8.0–10.0	8.2–9.2	8.8–9.2	5.0–8.3	—	—	—
Color (units)	—	—	—	—	—	—	—	—
Organics:								
MBAS	—	—	—	—	—	—	1	—
CCl$_4$	5	5	—	—	—	—	1	—
COD	<0.03	<0.03	0.5	0	75	75	75	75
Dissolved oxygen	—	—	<0.03	<0.005	—	—	—	—
Temperature, °F	120	120	120	120	100	120	100	120
Turbidity (JTU)	19	5	0.5	0.05	5,000	100	—	—

*Determined by treatment of other constituents.
SOURCE: Table 14-11 in Ref. 8.

TABLE 4.2 Water Quality Requirements (mg/L) Are Most Stringent for High-Pressure Boiler Feedwater; Requirements for Many Other Uses Can Be Achieved by a Well-Run Secondary Treatment Plant (Continued)

				Process water by industry				
Characteristics	Textile	Lumber	Pulp & paper	Chem.	Petro-leum & coal products	Prim. metals	Food canning	Bottled and canned soft drinks	Tanning
Silica (SiO_2)	—	—	50	50	60	—	50	—	—
Aluminum (Al)	—	—	—	—	—	—	—	—	—
Iron (Fe)	0.1	—	0.3	0.1	1.0	—	0.2	0.3	50
Manganese (Mn)	0.01	—	0.1	0.1	—	—	0.2	0.05	0.2
Copper (Cu)	0.05	—	—	—	—	—	—	—	—
Calcium (Ca)	—	—	20	70	75	—	100	—	60
Magnesium (Mg)	—	—	12	20	30	—	—	—	—
Sodium and potassium	—	—	—	—	230	—	—	—	—
Ammonia (NH_3)	—	—	—	—	40	—	—	—	—
Bicarbonate (HCO_3)	—	—	—	130	480	—	—	—	—
Sulfate (SO_4)	—	—	—	100	600	—	250	500	250
Chloride (Cl)	—	—	200	500	300	500	250	500	250
Fluoride (F)	—	—	—	5	1.2	—	1	1.7	—
Nitrate (NO_3)	—	—	—	—	10	—	10	—	—
Phosphate (PO_4)	—	—	—	—	—	—	—	—	—
Dissolved solids	100	—	100	1,000	1,000	1,500	500	—	—
Suspended solids	5	<3 mm in dia.	10	5	10	3,000	10	—	—
Hardness ($CaCO_3$)	25	—	475	250	350	1,000	250	—	150
Alkalinity ($CaCO_3$)	—	—	—	125	500	200	250	85	—
Acidity ($CaCO_3$)	—	—	—	—	—	75	—	—	—
pH (units)	6.0–8.0	5.0–9.0	4.6–9.4	5.5–9.0	6.0–9.0	5.0–9.0	6.5–8.5	—	6.0–8.0
Color (units)	5	—	10	20	25	—	5	10	5
Organics:									
MBAS	—	—	—	—	—	—	—	—	—
CCl_4	—	—	—	—	—	30	—	—	—
COD	—	—	—	—	—	—	—	—	—
Dissolved oxygen	—	—	—	—	—	100	—	—	—
Temperature, °F	—	—	100	—	—	—	—	—	—
Turbidity (JTU)	—	—	—	—	—	—	—	—	0

SOURCE: Table 14-11 in Ref. 8.

TABLE 4.3 Suggested Water Quality Limits for Industrial, High-Duty, Primary-Fuel, Drum-Type Boilers

Makeup water percentage: Up to 100% of feedwater
Conditions: No superheater, turbine drives, or process restriction on steam purity, but including deaerator

	Drum operating pressure	
	0–300 psig (0–2.07 MPa)	301–600 psig (2.08–4.14 MPa)
Feedwater:*		
Dissolved oxygen, mg/L O_2	<0.007	<0.007
Total iron, mg/L Fe	<0.1	<0.05
Total copper, mg/L Cu	<0.05	<0.025
Total hardness, mg/L $CaCO_3$	0.5	0.03
pH @ 25°C	8.3–10.5	8.3–10.5
Nonvolatile TOC, mg/L C†	<1	<1
Oily matter, mg/L	<1	<1
Boiler water:		
Silica, mg/L SiO_2	<150	<90
Total alkalinity, mg/L $CaCO_3$	<1000‡	<850‡
Specific conductance, µS/cm, without neutralization	<8000‡	<6500‡

*Boilers with rather large furnaces, large steam release space, and internal chelant, polymer, and/or antifoam treatment can often tolerate higher levels of feedwater impurities than those in the above table and still achieve adequate deposition control and steam purity. Removal of these impurities by external pretreatment is always a more positive solution. Alternatives must be evaluated as to practicality and economics in each individual case. The use of some dispersant and antifoam internal treatment is typical in this type of boiler operation, which allows the higher feedwater hardness.

†Nonvolatile TOC is that organic carbon not intentionally added as part of the water treatment.

‡Alkalinity and conductance values consistent with steam purity limit. Practical limits above or below the tabulated values should be individually established by careful steam purity testing.

SOURCE: Ref. 9.

boiler tube ruptures (Fig. 4.2). Installation of a water softener* (cation exchanger)[12] is usually necessary; this softener shall consist of a tank holding several cubic feet of a cross-linked sulfonated polystyrene resin. The commonly installed resin has a capacity of 28,000 grains of hardness (as calcium carbonate) per cubic foot. [1 gr/gal hardness (as $CaCO_3$) = 17.12 mg/L hardness (as $CaCO_3$)].

---

*These softeners were first called sodium zeolite softeners, as they were filled with a mineral formed from the sodium salt of hydrated aluminum silicate. These cation exchange softeners (regenerated with sodium chloride) are still called zeolite softeners, although they have a much lower capacity than softeners containing the presently manufactured polystyrene resins.

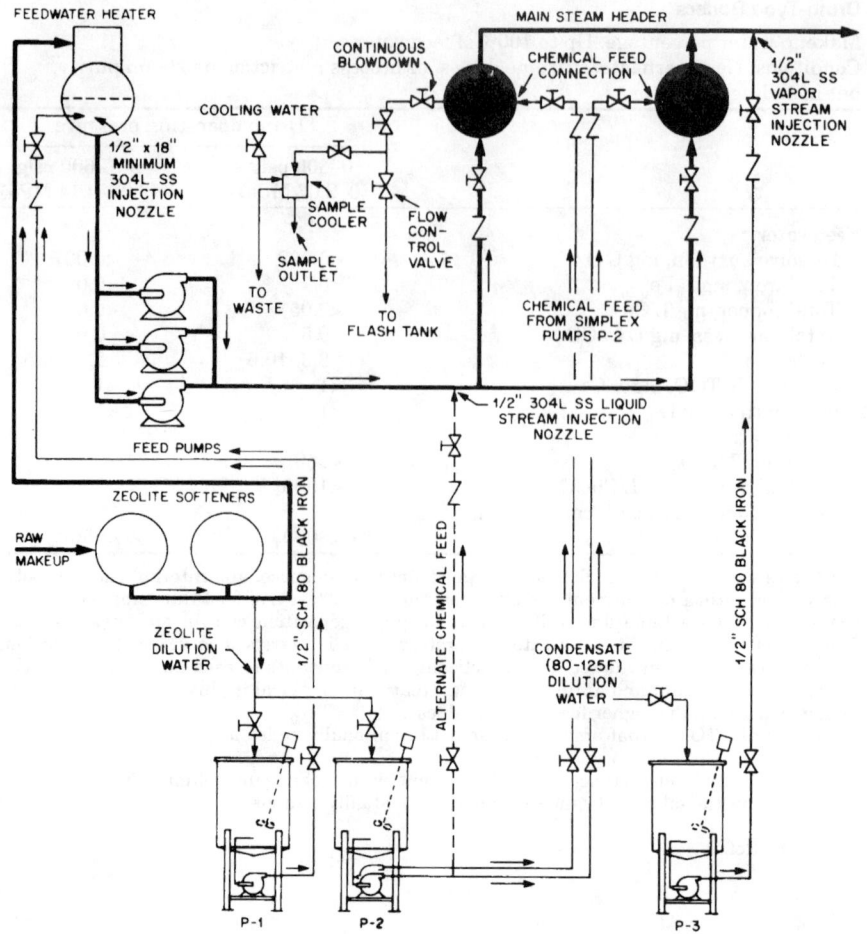

Figure 4.1 Typical piping for boiler system chemical control. (*Betz Laboratories, Trevose, Pa.*)

Calculations reveal that 100 ft³ of this resin treating water of 100 mg/L hardness content can soften:

$$28{,}000 \times 17.12 \times 100 \text{ ft}^3/100 \text{ mg/L hardness} = 479{,}000 \text{ gal}$$

This is based on using 14 lb of salt (NaCl) per 1 ft³ of resin for regeneration. The calcium and magnesium ions in the water, which represent the hardness content, are exchanged for sodium in passage through the softener. Sodium salts are soluble and therefore do not form scale or

Steam and Hot Water Heating Systems 77

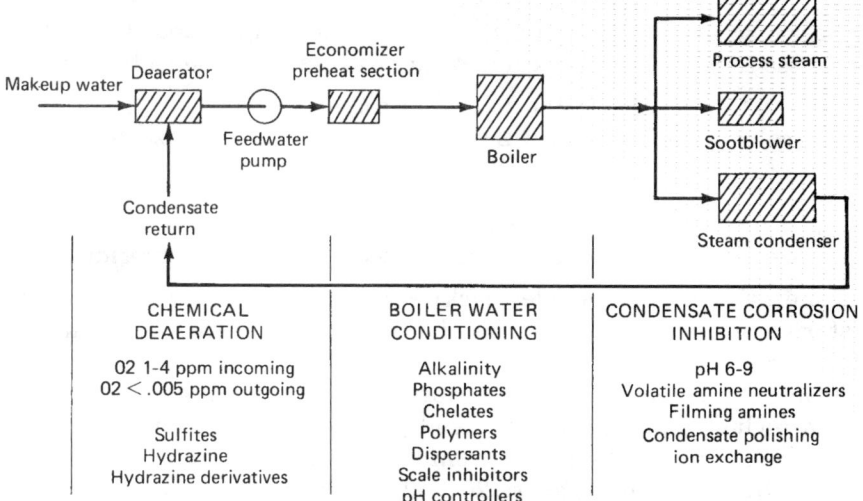

Figure 4.2 Key boiler-treatment chemicals. [*T. C. Elliott (ed.), Standard Handbook of Powerplant Engineering, McGraw-Hill, New York, 1989.*]

sludge. These ion exchangers (see Fig. 4.3) are usually sized so that regeneration with a sodium chloride (NaCl) brine solution is required every few days or weekly when effluent tests indicate that the capacity for removing hardness is exhausted; this is shown by a positive value from the hardness test of the softener effluent water. (See App. C for a

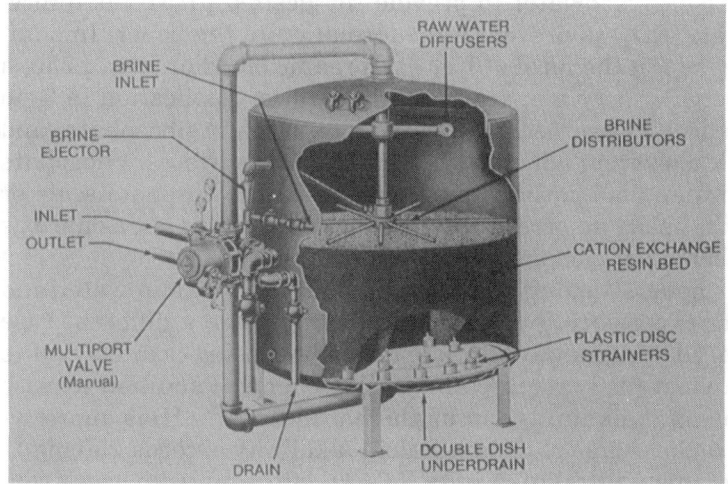

Figure 4.3 Cross section of an ion exchange unit. [*Frank N. Kemmer (ed.), The NALCO Water Handbook, 2d ed., McGraw-Hill, New York, 1988.*]

description of the hardness test.) These ion exchangers may be designed for periodic automatic regeneration. Chemical equations illustrating hardness removal and the regeneration with salt are shown below:

$$Ca(HCO_3)_2 + 2NaZ \rightarrow CaZ_2 + 2NaHCO_3$$
calcium bicarbonate + 2 sodium zeolite → calcium zeolite + 2 sodium bicarbonate

$$CaZ_2 + 2NaCl \rightarrow 2NaZ + CaCl_2$$
calcium zeolite + salt → sodium zeolite + calcium chloride

Regeneration, which includes backwashing, brining, slow rinsing, and fast rinsing, is required before the softener is placed back in service.[10-12]

From the makeup system, the water passes to the preboiler system, which includes feedline piping, feed pumps, and an economizer. The economizer, by absorbing heat from the flue gases, increases boiler efficiency by 2 percent for every 25°F (14°C) increase in feedwater temperature. Scale formation should not be a problem in these areas, provided that hardness is kept below 1 mg/L by proper ion exchange softener control; however, the continuous application of caustic soda may be necessary to raise the pH above 9.0 to inhibit corrosion.

Postchemical boiler water treatment for both low- and high-pressure boilers usually consists of the proportionate application of a dispersant, caustic soda, and phosphate and strict adherence to boiler water test limits, including blowdown control.

Specifically, postchemical treatment usually consists of Item 6 disodium phosphate and Item 4 caustic soda (see App. A), which are applied to the boiler feedwater to provide the desired pH (P alkalinity) and phosphate (PO_4) to prevent scale formation in the boiler. In addition, dispersants in the form of Item 1F organic blend or Item L236, a modern copolymer acrylate, are recommended for application in 3- to 10-mg/L dosages to keep calcium carbonate, calcium phosphate, and iron oxide sludges from adhering to heat transfer surfaces. Phosphate is applied rather than carbonate because calcium phosphate scale or sludge is more easily dispersed and because this scale is not as hard as calcium carbonate or calcium sulfate scale.

Hot water boilers[13] using a minimal amount of makeup water and therefore not experiencing a scaling problem require a different type of postchemical treatment. Also, steam boilers using sodium zeolite make-up may not likely require an alkaline chemical addition to raise the pH if the total alkalinity minus the hardness (M − H) is appreciable and the boiler water phenolphthalein alkalinity exceeds 250 mg/L, as a result of this natural alkalinity (M − H) of the makeup water concentrating in the boiler water.

Prevention of corrosion

Application of Item 10 sodium sulfite is necessary in water treatment for steam boilers to remove traces of dissolved oxygen left after deaeration. Such traces would cause corrosion of the boiler itself and subsequent contacts with steam condensate. Facilities using makeup water of high alkalinity (>50 ppm) may experience acidic corrosion in return condensate systems caused by low-pH conditions resulting from carbon dioxide. Carbon dioxide in the condensate returns is the result of the breakdown of the natural bicarbonate content into carbonate and hydroxide when makeup water is subjected to boiler water temperatures. In this case, amines such as Item 13 cyclohexylamine, Item 30 morpholine, and/or Item 116 diethylethanolamine are specified to be applied continuously (not by slug feed or pot feeder) with the softening chemicals to the boiler feedwater. Sufficient amines should be applied to provide a pH of at least 7.5 and an added total (methyl orange) alkalinity of 10 to 30 mg/L in the return condensate.

Humidification. Overfeed of amines should be avoided, since respiratory problems may result if steam is applied for room humidification.[14] With regard to designing steam humidification systems,[15] it is now pointed out that chemically contaminated steam should not be installed except in cases in which no other method is available. While OSHA/ACGIH (Occupational Safety and Health Administration/ American Conference of Governmental and Industrial Hygienists) limits and guidelines are not strict, it is emphasized that the public's right-to-know law is in force. Data released to date do not indicate serious contamination;[15] however, the effect of habitual contact of workers in an area of low-level amines on a long-term basis hasn't been determined. While Material Safety Data Sheet (MSDS) information is provided by the amine supplier, it is questioned whether the worker or occasional hospital patient is made aware of the contamination (the right to know).

The number of air changes and amine dosage levels will be expected to have considerable impact on the level of air contamination. Heat-exchanger-produced steam from corrosion-resistant boilers is one alternative source of humidification. Another is the addition of water mists to the conditioned airflow, but it should be recognized that stagnant water pools may develop bacterial growth (pathogens) and must be avoided. It is essential that water used for humidification have minimal dissolved solids and scale-forming substances; otherwise, deposition may be expected.

In cases of high makeup water usage and high makeup water alkalinity, installation of a dealkalizer (an anion exchanger regenerated with salt or a cation exchanger that contains a carboxylic resin and is regenerated with acid), in addition to the sodium zeolite softener, may be required. The dealkalizer reduces the bicarbonate alkalinity (potential carbon dioxide) of the makeup water to 5 to 15 mg/L and reduces the corrosion resulting from carbon dioxide. Another advantage is that it reduces the amine treatment requirements and provides a more corrosion-resistant return condensate system.

Deaerator. The *deaerator* (Fig. 4.4) fills an important role[9,10] in lowering the dissolved oxygen content from about 10 mg/L to 0.03 mg/L. In most modern types of deaerators, the feedwater falls in droplets through rising steam under pressure, and the air (dissolved oxygen) is vented out the top of the deaerator. To ensure that this equipment is functioning properly, one can check the steam pressure in the deaerator against the effluent temperature and determine if undesired air binding is occurring in the heater. Table 4.4 shows the corresponding pressure and temperature readings to be expected in the feedwater heater. The pressure and temperature should correspond; for example, the temperature should be 227°F at 5 psig, but if it is 222°F, then air binding is indicated. The pressure and temperature should be checked regularly to assure that water flow is not being partially interrupted by

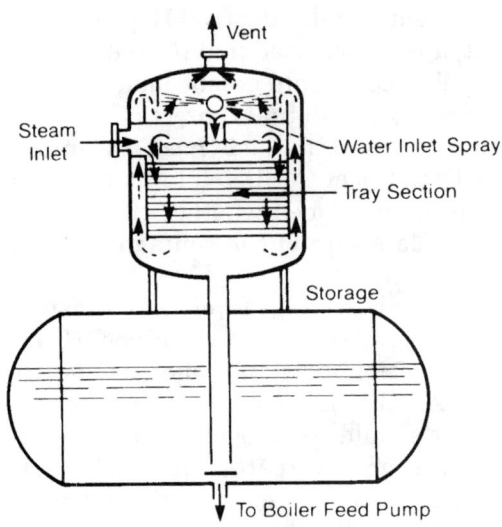

Figure 4.4 (*a*) Spray-tray deaerator. [*Frank N. Kemmer (ed.), The NALCO Water Handbook, 2d ed., McGraw-Hill, New York, 1988.*]

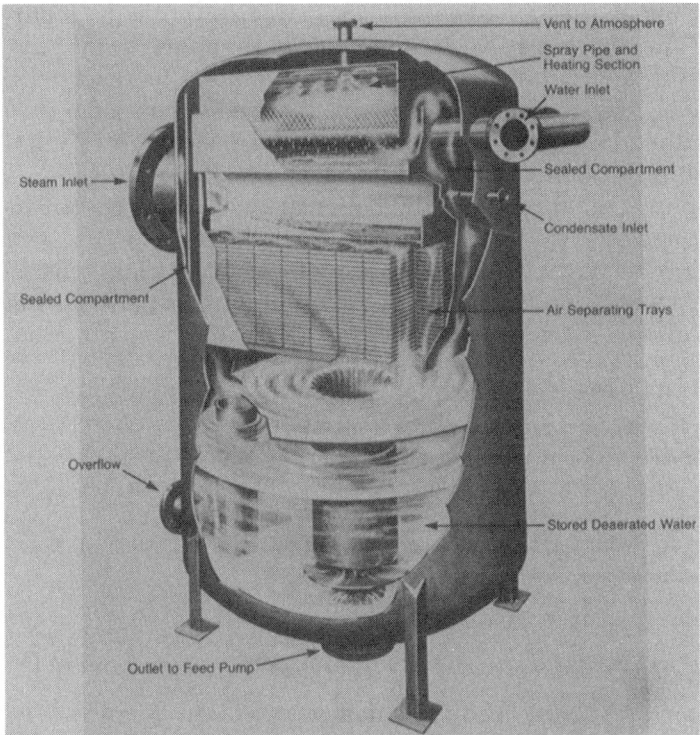

Figure 4.4 (b) Tray-type deaerating heater. [*Frank N. Kemmer (ed.), The NALCO Water Handbook, 2d ed., McGraw-Hill, New York, 1988.*]

TABLE 4.4 Corresponding Pressures and Temperatures to Be Expected in Feedwater Heaters

Feedwater heater steam gage, psig	Feedwater heater temperature	
	°F	°C
0	212	100
1	215.3	101.8
2	218.5	103.6
3	221.5	105.3
4	224.4	106.9
5	227.1	108.4
6	229.8	109.9
7	232.3	111.3
8	234.8	112.7
9	237.1	113.9
10	239.4	115.2

heater tray scaling or that air release is not being prevented by a partial stoppage of the top vent valve. Sodium sulfite is applied to the storage section of the deaerator or the boiler feed pumps to react with residual dissolved oxygen not removed in the feedwater deaerator. Sufficient caustic soda is also applied to attain the desired pH (P alkalinity) specified in the control limits. Item L236 dispersant and Item 6 disodium phosphate are applied to the boiler drum to meet the treatment specifications and limits previously prescribed.

While there are many continuous-recording instruments for measuring the dissolved oxygen in a deaerator effluent, an inexpensive way of estimating the dissolved oxygen is to measure the sulfite consumption.

Example Data from power plant log:

Sulfite (SO_3) = 45 mg/L (average of boiler water tests)

COC (cycles of concentration) (calculated from feedwater and boiler chloride conductivity tests): boiler water, 2500 µS/cm conductivity, and feedwater, 140 µS/cm conductivity; then COC = 2500/140 = 17.8

FW (feedwater) or steam produced per day = 1,017,000 lb

Sodium sulfite *dosage* per day = 5.5 lb (from daily log)

Calculation of dissolved oxygen (DO) in feedwater:

$$DO = \{[dosage \times (1,000,000/FW)] - [(SO_3 \times 1.574)/COC]\} \times 0.127 = 0.18 \text{ mg/L}$$

Another concern involving feedwater deaerators is their safety.[16] In recent years, particularly in the pulp and paper industry, there have been cases of serious storage vessel ruptures resulting from deaerator weld cracking. Nondestructive examination (NDE) of welds, both in the heater and in the storage sections, is recommended. Water hammer and vibration incidences may have been instrumental in the failures.[16]

Figure 4.5 shows the solubility of dissolved oxygen in water at different temperatures.

Prevention of stress corrosion cracking

Years ago, stress corrosion cracking, or caustic embrittlement, was a serious problem in boilers because excessive alkaline treatment of the boiler water was allowed to accumulate in the riveted seams, which eventually caused stress corrosion cracking, metal ruptures, and boiler explosions. The cold-rolled tube ends in modern boilers are usually the only parts now subject to this attack. Since sulfite converts to sulfate in reaction with dissolved oxygen, the high sulfite dosages likely required in systems not including a deaerator may probably provide sufficient sulfate to meet the specified requirement of a 1:1 ratio of sodium sulfate to alkalinity [as sodium carbonate (Na_2CO_3)] in boiler water. This ratio is necessary to inhibit caustic embrittlement. Other corro-

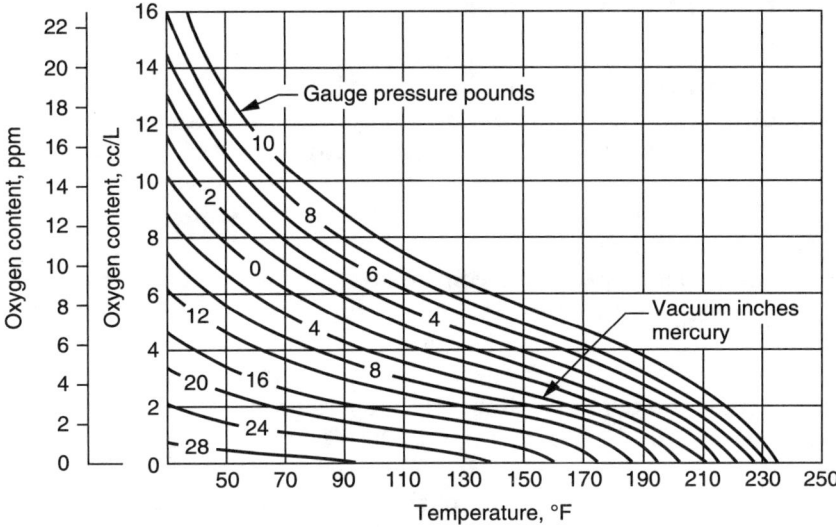

Figure 4.5 Solubility of dissolved oxygen. (*R. W. Lane files.*)

sion-inhibiting treatments for this purpose include the maintenance of 40 mg/L Item 9 sodium nitrate (as NO_3) and 100 mg/L Item 1F organic blend in the boiler water.

Prevention of boiler water carryover

Keeping boiler water dissolved solids within the prescribed test limits is the most effective method of preventing carryover from steam boilers. Dispersant formulations, such as Items 1F and L236, which contain antifoams, are also helpful in preventing carryover. In power plants producing electricity in addition to thermal energy, steam produced may be superheated in order to increase steam turbine efficiency and to reduce turbine corrosion. For example, the temperature of 250 psig saturated steam may be increased from 406 to 506°F (208 to 263°C) by heating the saturated steam in a superheater. This steam may be extracted from an intermediate stage of the turbine to supply process steam demands and may retain appreciable superheat; therefore, the addition of pure water (return condensate or feedwater) is required to restore it to saturated-steam conditions. Saturated steam is preferred for comfort heating or process steam uses owing to its ready availability of the heat of vaporization on condensing to water in heat exchangers. Water quality is of prime concern in the desuperheater, since poor-quality feedwater used for desuperheating can contaminate the steam, and impure steam could cause deposition and corrosion.

Proper Water and Steam Sampling

Accurate water samples from high-pressure boilers are best obtained from the continuous-blowdown line after the water has been cooled to room temperature in a sample cooler. A sample cooler usually consists of a small counterflow heat exchanger containing a stainless-steel coil; the boiler water is cooled in passage through the heat exchanger.

In sampling, one must first turn on the cooling water, then run the sample to waste for at least 30 s, and then adjust the effluent valve of the sample cooler to provide a continuous flow of about 1 pint of sample per minute and an effluent temperature below 120°F (49°C). The sample container should then be rinsed with the sample at least twice before the sample is taken for analysis. The sulfite test (App. C) should be conducted immediately, since air may react with the sample, depleting the sulfite. Water tests should be conducted daily or during each shift and the results recorded for hardness, P alkalinity, M alkalinity, chloride, and conductivity of the makeup water, feedwater, and condensate water according to the analytical methods described in App. C. All power plants should provide a small laboratory space that is clean, pleasant, efficient, and adequately equipped for conveniently conducting water tests (see Fig. 4.6). This was part of the reason for the

Figure 4.6 Power plant water-testing laboratory. (*R. W. Lane files.*)

successful launching of a water treatment program in the state of Illinois. Actually, there was competition between the different chief engineers to see which one could install the most efficient testing laboratory; as a result, the operating engineers found these pleasant surroundings conducive to frequent and accurate water testing.

Immediate action should be taken to correct operations that may be responsible for high-hardness and off-limit test results for alkalinity, chloride, and conductivity. Such action should be taken before serious effects are observed in the boiler.

Obtaining a representative water sample for testing is extremely important, for trying to correct for a supposed deficiency indicated by a nonrepresentative sample could lead to serious maintenance problems. While the continuous-blowdown water provides an accurate sample, the only sampling location available in small low-pressure steam and hot water boilers may be the water column. As the samples being taken can be very hot, a sample cooler, which cools the sample by passage through a small heat exchanger cooled by cold water, should be installed at each location so that water samples below 120°F (49°C) can be obtained for accurate water testing. Sampling from the water column requires the following special care:

1. Since the water column contains stagnant water, it must be blown down completely at least twice in order to obtain fresh water from the boiler itself.
2. In some boilers, the automatic makeup valve may add water at the water column, so this must closed; otherwise, makeup water rather than boiler water will be sampled. After the boiler water is sampled, *the makeup valve must be reopened.*
3. The sample should be taken rather slowly; if sampling is done too rapidly, steam rather than boiler water will be sampled. The water level in the water column should be observed closely as one samples to make sure that boiler water is being sampled, not the steam that is above the boiler water.

The modern high-pressure power plant requires accurate sampling from numerous locations for proper monitoring and satisfactory control in order to avoid possible scaling or corrosive tendencies. Continuous analyzers are often necessary. This means that careful consideration must be given to the location, size, and material of the sampling line and to the flow rate and length of the sampling line. Steam should be fully condensed close to the boiler to avoid superheating of the sample and deposition of contaminants in the sampling line. The reason for taking this precaution is that boiler water solids such as sodium sulfate and sodium chloride are more soluble in saturated steam than in superheated steam. When steam is sampled in a long line and pressure is gradually reduced, it becomes superheated, and the salts become less soluble and deposit in the sampling line, yielding an inaccurate sample. The sampling line should be of stainless steel, and the flow rate should be at least 6 ft/s.[17] Advice on proper sampling is covered in detail in ASTM D3370.

Chemical Feed Systems

It is essential that reliable chemical feed systems be installed to ensure continuous proportional feeding of needed chemicals to locations where optimum benefit is derived. This may likely mean that proportional feeding of the necessary chemical may be actuated by a water meter or possibly a pH meter in order to provide accurate chemical effectiveness. Modern high-pressure boilers require that external treatment systems be installed to provide minimum hardness, bicarbonate alkalinity, silica, and dissolved solids (specific conductivity). Ion exchange and even reverse osmosis and demineralization are now being justified for the lower pressures (100 psig); however, at the higher boiler pressures (>250 to 400 psig), such installations are considered a "must" to ensure minimum scale and corrosion. Of course, the consultant, in deciding on the proper overall water treatment method, considers the cost of these external treatment systems, but he or she must also consider the cost of increased blowdown, as well as costs for internal water treatment without proper external treatment. Exact proportional feed may be needed in cases in which several waters of different quality are in use and in cases of diverse loads and boiler capacities. However, continuous application of postchemical treatment to the common boiler feed-pump effluent over a 24-h period usually provides adequate proportionate feed. Essentially, this means that the boiler operating at highest capacity will take more feedwater and therefore receive more treatment. It is assumed, of course, that the daily chemical charge added per chemical tank is adjusted properly to attain the prescribed alkalinity, phosphate, sulfite, and dissolved solids.

A complete internal treatment system is illustrated in Fig. 4.1. Chemical solutions are fed typically by 1-gal/h corrosion-resistant pumps to pressurized (100 to 700 psig) locations from 50- to 100-gal vats equipped with electric mixing equipment. Sodium ion exchange–treated water or steam condensate is recommended for dissolving the chemicals, and daily vat charges are usually based on the addition of adequate chemicals for the day's requirement. The purpose of water testing is to determine the boiler water constituents, maintain them at the specified level, and adjust the vat charge to attain this level of treatment.

Assuming that water of minimal hardness is being applied as makeup, the usual treatment plan is as follows:

1. Sufficient Item 4 caustic soda to raise the feedwater pH to approximately 9.0 is applied to reduce corrosive tendencies in the feed lines and economizer.
2. Next, an oxygen scavenger, in the form of Item 10 sodium sulfite, is applied to the storage section of the deaerating feedwater heater or to the boiler feed-pump effluent to complete the removal of dissolved oxygen, which could cause system corrosion.
3. Item 4 caustic soda and a phosphate, such as Item 6 disodium phosphate, are usually applied with the sulfite or could be applied directly to the boiler drum. The advantage of applying these chemicals to the feedwater rather than to the individual boiler is that the different boilers get appropriate quantities on the basis of their immediate particular load. In addition, previously discussed amine treatment, such as Item 13 cyclohexylamine and Item 30 morpholine, may be required to neutralize acidic steam condensate. The required amine treatment may be applied with the sulfite, caustic soda, and phosphate chemicals. The amount of chemicals applied to the treatment tank should be based on the recommended test levels prescribed in the control chart for boiler feedwater treatment (see Table 4.5 for an example of a control chart).

Continuous Blowdown

Continuous flow of blowdown has proved to be a very satisfactory method of maintaining accurate control of dissolved solids in boiler water within the prescribed limits. To attain continuous blowdown in steam boilers (Fig. 4.7), it is necessary to install a pipe opening 1 to 4 in below the normal water level inside the boiler and then to extend the pipe to a metering valve outside the boiler. Adjustment of this valve provides the degree of blowdown desired. In large systems, recovery of this wasted heat (Fig. 4.8) is justified and involves the installation of a heat exchanger in which the incoming makeup water is heated by the hot blowdown water before passage to the boiler. Manual blowdown of other valves on steam boilers is necessary mainly to relieve localized sludge accumulations. These valves may need to be opened only for brief periods once a day to twice a week.

TABLE 4.5 Typical Boiler Feedwater Control Chart

Name of Facility
Chicago, Ill.   Date
Boiler Feedwater Control Chart
Zeolite Softening plus Dealkalization of Chicago City Water

Tests	Sample size, mL	Recommended		
		mL		mg/L
Dealkalizer:				
Hardness (H)	50	0.0		0.0
M alkalinity (M)	50	0.5–1.5		10–30
Boiler:				
P alkalinity (P)	20	6.0–9.0	× 20 =	300–450
Phosphate (PO$_4$)				20–50
Sulfite (SO$_3$)	50	1.0–2.0	× 20 =	20–40
Conductivity				3000–4000 μS/cm
Condensate:				
Hardness (H)	100			0.0
P alkalinity (P)	100			Trace
M alkalinity (M)		0.5–3.0	× 10 =	5–30

Control. Makeup in the form of completely softened zeolite water shall be passed through a dealkalizer regenerated with 6 lb salt and 1/2 lb Item 4A liquid caustic soda/ft^3. The postchemical feed system shall apply Item L240A dispersant, Item 4A liquid caustic soda, Item 23 sodium tripolyphosphate, Item 10 sodium sulfite, and amine mixture to the pressure side of the boiler feed pumps.

Organic blend. 4 lb Item L240A/million lb steam produced is recommended.

P alkalinity. Apply sufficient caustic soda to maintain the P alkalinity of 300 to 450 mg/L in the boiler water.

Phosphate. If the phosphate tests below 20 mg/L, increase phosphate. If the phosphate tests above 50 mg/L, reduce phosphate.

Sulfite. If the sulfite tests below 20 mg/L, increase sulfite. If the sulfite tests above 40 mg/L, reduce sulfite.

Condensate treatment. A 50:50 mixture of Item 13 cyclohexylamine and Item 30 morpholine shall be applied to provide a trace P alkalinity (pH 8.2) and an M alkalinity of 10–30 mg/L.

Blowdown. If conductivity exceeds 4000 μS/cm, increase blowdown. If conductivity is less than 3000 μS/cm, decrease blowdown. Continuous blowdown is to provide the main means of controlling boiler solids; however, a minimum of one "flash" blowdown shall be applied to all blowdown valves once to three times weekly. Additional manual blowdowns shall be applied according to the amount of scale or sludge observed during boiler inspections.

Signed, Water Treatment Consultant

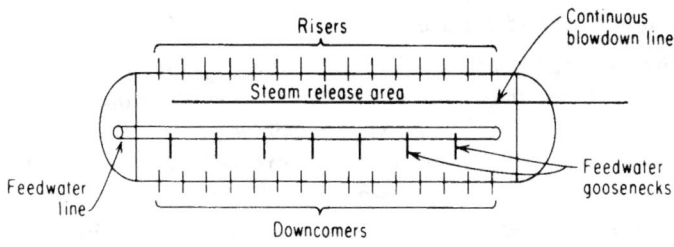

(a) Top view of boiler drum

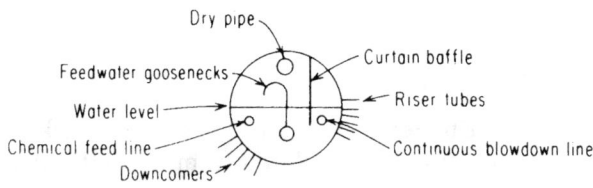

(b) End view of boiler drum

Figure 4.7 Typical continuous blowdown location. [*Frank N. Kemmer (ed.), The NALCO Water Handbook, 2d ed., McGraw-Hill, New York, 1988.*]

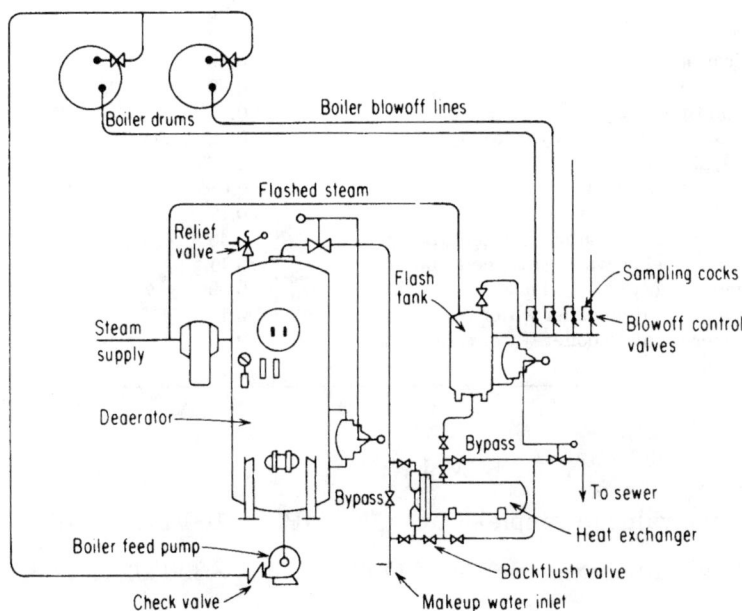

Figure 4.8 Heat recovery from continuous blowdown. [*Frank N. Kemmer (ed.), The NALCO Water Handbook, McGraw-Hill, New York, 1988.*]

Hot water boiler treatment and control[13]

Provided that the water in hot water boilers is properly treated and that steam or vapor is not lost from the system and concentration does not occur, hot water boilers should practically never be blown down. Since the water in these systems has had the oxygen used up by the initial corrosion occurring in these systems, replacing this original water with fresh water containing oxygen only promotes further corrosion by the added oxygen. Therefore, these boilers should not be blown down and fresh water should not be added for replacement; this is one of the secrets to their general maintenance-free operation.

Discussion of Boiler Water Test Limits

Estimates of the required boiler feedwater dosages for package-type treatment of a hard water may be calculated from the following information and equations:

Treatment (Item no.)	Softening value	Phosphate value
4, caustic soda	1.25	
5, soda ash	0.94	
sodium bicarbonate	0.59	
potassium hydroxide	0.89	
6, disodium phosphate	0.70	1.06
7, trisodium phosphate, monohydrate	0.82	0.82
8, sodium polyphosphate	0.58	1.35
23, sodium tripolyphosphate	0.68	1.22
119, softened water boiler treatment	0.12	
148, low-pressure boiler water treatment, powdered	0.48	
149, low-pressure boiler water treatment, liquid	0.11	
L219, copolymer-phosphate boiler water treatment	0.09	
L236, copolymer-alkaline boiler water treatment	0.13	
L240, copolymer-alkaline boiler water treatment, FDA approved	0.12	

$$\text{Mol Wt}$$

$$\text{Softening value (example): } CaCO_3/Na_2HPO_4 = 100/142 = 0.70$$

$$\text{Phosphate value (example): } 3CaCO_3/2Na_2HPO_4 = 300/284 = 1.06$$

The softening value may be defined as the $CaCO_3$ alkalinity provided per mole, and the phosphate value may be defined as the alkalinity provided in forming $Ca_3(PO_4)_2$. If both the analysis of the makeup water and the softening value of the treatment are known, the approximate quantity of treatment required may be calculated as follows:

Pounds/1000 gal = (hardness − M alkalinity) + 40/(softening value)(120)

An example would be as follows: H = 140 mg/L; M = 110 mg/L; softening value = 0.13; (140 − 110) + 40 = 70 mg/L required; then 70/(0.13 × 120) = 4.5 lb/1000 gal Item 236.

In the case of softened makeup water, the natural alkalinity (M alkalinity) of the water would provide 110 mg/L feedwater alkalinity and, on the basis of 10 boiler water concentrations, an estimated phenolphthalein alkalinity (75 percent of M alkalinity) of 825 mg/L and 1100 mg/L M alkalinity. So there is no need to add alkalinity (caustic soda) to obtain the desired alkalinity in the boiler. Item 6 disodium phosphate requirements to obtain 35 mg/L PO_4 in the boiler water can be calculated by assuming a softener effluent of 0.2 mg/L hardness plus that needed as shown below:

Pounds Item 6 required = 0.2/phosphate value + 3.5/phosphate value

Example: 0.2/1.06 + 3.5/1.06 = 3.5 mg/L/120 = 0.03 lb/1000 gal Item 6.

Control values for feedwater and boiler water quality are given in Table 4.6.

TABLE 4.6 Control Values for Feedwater and Boiler Water Quality

Tests	Limits, mg/L or ppm*
P alkalinity (as $CaCO_3$)	250–500†
Phosphate (PO_4)	30–60
Sulfite (SO_3)	30–50
Conductivity	2500–4000 µS/cm‡

*These normal *test limits* are prescribed in steam boiler water treatment control charts provided for each specific boiler installation.

†When zeolite softening is applied to a high-alkalinity makeup water, the limit prescribed will be higher (as high as 1000 mg/L P alkalinity).

‡These limits, usually based on maintaining 20 to 50 feedwater cycles, may vary depending on the equipment, boiler pressure, required capacities, and services.

SOURCE: American Society of Mechanical Engineers, New York, 1990.

Demineralization or Dealkalization

The high-pressure electric utility plants use complete demineralization systems to provide makeup water of minimal dissolved solids and carbonate alkalinity (potential carbon dioxide) content.

An industrial or building water treatment facility cannot usually justify the expense of these complex systems and may not have qualified personnel to operate them. However, installation of a dealkalizer to supplement the sodium zeolite softener and to remove the bicarbonate alkalinity is justified in many cases in which the raw water alkalinity is appreciable (>50 mg/L M alkalinity). Such equipment is not expensive; it costs somewhat more than a sodium zeolite softener. In its simplest form it consists of an anion resin exchanger, which is regenerated with sodium chloride and converts the makeup water alkalinity to the chloride form in passage through the dealkalizer following the sodium ion exchanger.

An example of the softener and dealkalizer sizing required for a small plant that produces 100,000 lb of steam a day at 50% makeup and employs moderately hard water [150 mg/L (as $CaCO_3$) equivalent to 8.8 gr/gal hardness, and 140 mg/L total alkalinity (as $CaCO_3$)] is as follows:

$$(100,000 \times 0.5)/8.34 = 6000 \text{ gal/day}; 6000/1440 = 4.2 \text{ gpm}$$

So decide on a maximum flow of 10 gpm. On the basis of 8.8 gr/gal hardness, for 2 days' operation, $8.8 \times 6000 \times 2 = 100,000$ gr of softener required or $100,000/28,000$ gr/ft^3 = 4 ft^3 of resin. Also, the dealkalizer (100,000 gr) would require a tank with a similar quantity of anion resin.

For large facilities and those having a makeup water of a high total (M) alkalinity (>100 to 150 mg/L), installation of a cation ion exchanger that contains a carboxylic-type resin and uses sulfuric acid regeneration will prove more economical. This ion exchanger provides superior-quality makeup water, as it actually reduces the dissolved solids content of the makeup water rather than increasing it, as salt-regenerated cation and anion exchange methods do.

Chelant treatment[18] is now being regularly prescribed as a substitute for part or all of the phosphate requirements because it provides cleaner boilers, free of deposits of calcium phosphate scale or sludge. Chelants have the desirable property of dissolving water hardness and keeping it in solution. The sodium salt of ethylenediamine tetracetic acid (EDTA) (Item 68), which is the most commonly used chelant, should be applied separately through a stainless-steel quill into the feedwater just before entering the boiler drum. Caustic soda and sodium sulfite should be applied to the storage section of the feedwater heater in order to render the feedwater completely free of dissolved oxygen. Dissolved oxygen degrades chelants, and so sulfite must be applied before the addition of chelant. Dispersant (Item 236) and a low level of phosphate* are applied directly to the boiler drum. Sufficient chelant is applied to provide a chelant excess of about 1 mg/L (as $CaCO_3$) in the feedwater sample taken just after the chelant application point. Tests for chelant in the boiler water are not informative for control purposes. *Overfeed of chelant should be avoided,* since overfeed can cause serious boiler drum corrosion.

Treatment for Condensate Return Systems

The commonly observed corrosion in condensate return systems discloses the need for accurate and conscientious control of corrosion in these systems. As discussed previously, the natural alkalinity of the makeup water decomposes to carbon dioxide when exposed to the heat of the boiler water. The result is that the carbon dioxide in the steam is dissolved in the return condensate; carbonic acid is formed, and this causes acidic corrosion.

To neutralize the carbonic acid, volatile amines (Item 30 morpholine, Item 13 cyclohexylamine, and/or Item 116 diethylethanolamine) are applied with the caustic soda, phosphate, and sulfite chemicals to the boiler feedwater. While this technique is fairly effective, the amines and carbon dioxide have different properties of volatility; as a result, the desired perfect neutralization may not be attained under all pressure and temperature conditions in a plant condensate system.

The properties of the various amines are listed in Table 4.7, and Fig. 4.9 shows the pH attained by different dosages of the different amines. The differences in the distribution ratios are particularly striking; the ratios indicate that ammonia is the most volatile and that AMP95 (95% 2-amino 2-methyl 1-propanol) is the least volatile. This property of distilling from boiler feedwater is important in deciding on the particular amine to apply to attain neutralization in a designated plant area of a certain pressure and temperature. Reference should be made to App. A to learn the percentage of the active ingredient in the commercial form of the amine being purchased.

High dosage levels of amines may be required for neutralization of high carbon dioxide levels (>30 ppm). Such dosages of amines (>10 ppm), which exceed FDA regulations, are a serious contaminant of

---

*Theoretically, phosphate is supposed to react with hardness before excess chelant, since the solubility product of calcium phosphate is higher than the hardness chelant complex. My experience however has not indicated that the presence of phosphate hinders the effectiveness of the chelant treatment and the phosphate test does provide additional assurance that adequate softening chemicals have been applied.

TABLE 4.7 Properties of Amines

	Ammonia (NH$_3$)	AMP95 (C$_4$H$_{11}$NO)	Cyclohex-ylamine (C$_6$H$_{13}$N)	DEAE (C$_6$H$_{15}$NO)	Morph-oline (C$_4$H$_9$NO)
Molecular weight	17	89	99	117	87
Boiling point, °F, at 1 atm pressure	—		273	325	264
Azeotrope:					
bp at 760 mmHg, °F	—		206	210	—
% amine in vapor			44	25	—
Basicity, pK$_b$, 25°C	4.75	4.30	3.8	4.28	5.62
Approx. pH at 0.001 N	—	—	10.5	10.3	9.8
Dosage required to adjust condensate pH to 7.4; ppm/ppm CO$_2$			2.0		1.93
Distribution ratio, vapor/liquid at atmospheric pressure	10.0	0.31 (@80 psig)	2.6–4.0	1.7	0.4
Open cup flash point, °F*			90	135	102

*Commercial preparations have higher flash points.

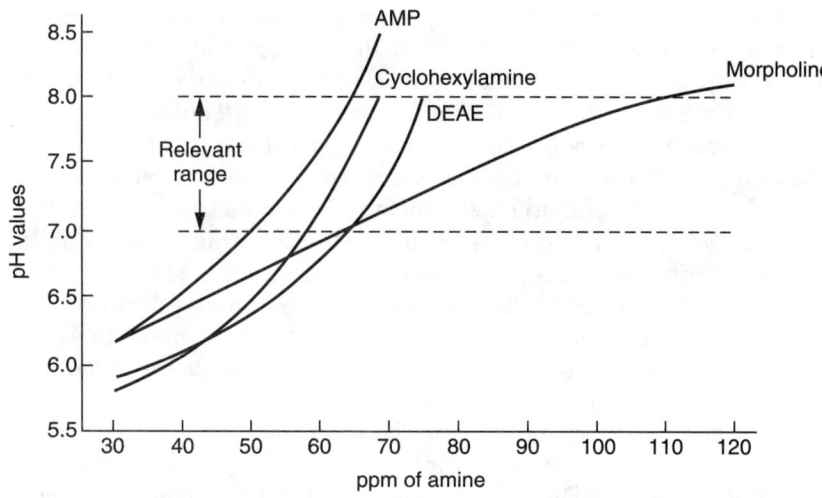

Figure 4.9 PPM of amines vs. pH. (*R. W. Lane files.*)

high-purity steam, as discussed above in the section on humidification. Some controversy still exists on this subject,[14,15] but recent studies have revealed that the amine concentrations are very low in a humidified room when compared with any established health standards.

Item 25 filming amine, an octadecylamine, is also used for inhibiting corrosion by forming a film in condensate return lines.[10] It is not generally recommended because a very exact dosage must be applied; otherwise, a slight excess causes accumulations and stoppages composed of an oily deposit of iron oxide in traps and equipment. In recent years, both filming and neutralizing amines and various combinations of the neutralizing amines have been applied to reduce corrosive conditions.

In addition, catalyzed hydrazine (Item 117) has been reported as being effective in reducing corrosion in condensate return lines and may be applied at several particularly corrosive locations in industrial plants covering a large area. Hydrazine has the advantage of reacting with oxygen and preventing pitting resulting from differential aeration and the presence of oxygen. This advantage is not provided materially by the neutralizing amines.

A number of different organic oxygen-scavenging agents are being marketed by the water treatment companies as substitutes for hydrazine, which is reported to be carcinogenic. Two of these, catalyzed diethylhydroxylamine (DEHA) and methylethyl ketoxime (MEKO), developed mainly for the high-pressure electric utility plants, have desirable properties for inhibiting corrosion by oxygen in condensate return systems of industrial plants. They are sufficiently volatile to distill over with the steam when applied to the deaerator effluent or steam boiler and should be beneficial in reducing pitting caused by oxygen in the condensate return systems. The author has not had any practical experience with these chemicals in industrial facilities but plans to try them out in condensate return systems experiencing pitting. Amine treatments only provide neutralization of carbonic acid; they do not provide inhibition of pitting by oxygen.

Operation and Proper Maintenance

This section of the chapter describes and stresses proper maintenance measures for efficient operation of (*a*) low-pressure steam and hot water boilers, (*b*) high-pressure steam boilers, and (*c*) electric boilers.

Low-pressure steam and hot water boilers

Low-pressure steam boilers are defined as boilers producing steam at pressures not above 15 psig (pounds per square inch gage). Hot water boilers produce only hot water and operate at pressures not above 160 psig and temperatures not above 250°C (482°F). These hot water boilers are usually constructed of steel, though some may be constructed of cast iron and are limited in pressure to 15 psig for steam and 30 psig for hot water. Usually these steam and hot water boilers[13] are gas- or oil-fired and range in capacity from 600 to 15,000 lb of steam (or equivalent of hot water) per hour. They are either the fire-tube type (Fig. 4.10), in which the hot flue gases pass through the boiler tubes, or the water-tube type, in which the hot flue gases pass on the outside of the tubes.

These small boilers, seemingly unimportant, generally do not receive adequate care with regard to maintaining clean fireside and waterside surfaces. As a result, many times their life may be only 2 to 5 years, while with proper maintenance, their life could easily be 20 years. When these boilers operate only in the winter season, it is particularly necessary to clean the soot deposits from the fireside surfaces and to spray these surfaces with oil to prevent acidic corrosion caused by soot and humid out-of-service conditions. The best advice for controlling corrosion of the boiler waterside surfaces during the summer outage is

Figure 4.10 Fire-tube boilers. (a) Two-pass fire-tube boiler. [*T. C. Elliott (ed.), Standard Handbook of Powerplant Engineering, McGraw-Hill, New York, 1989.*]

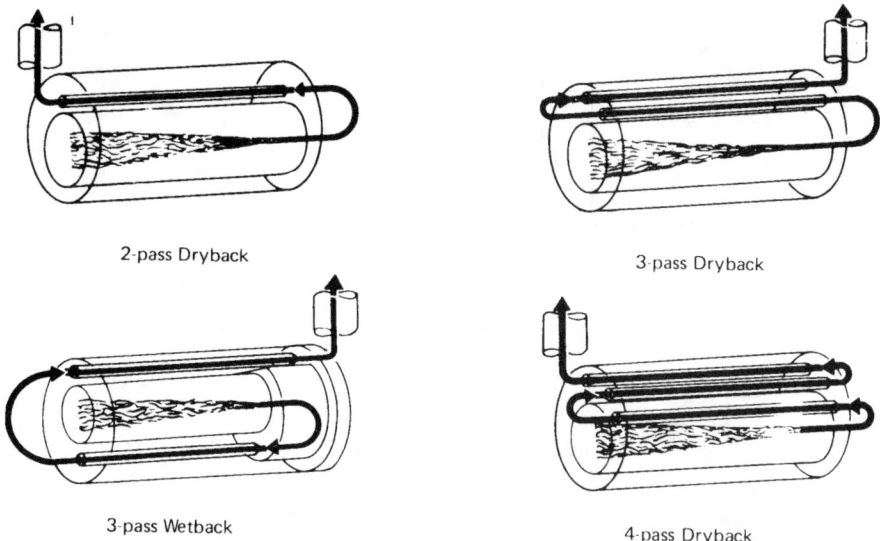

Figure 4.10 Fire-tube boilers. (*b*) Basic gas flow patterns used today in fire-tube boilers. [*T. C. Elliott (ed.), Standard Handbook of Powerplant Engineering, McGraw-Hill, New York, 1989.*]

not to drain the boiler. Keep it completely full of water after adding extra chemical treatment. This applies to low-pressure steam boilers as well as to hot water boilers, but in this case, extra chemical treatment is applied for the whole system and then is circulated throughout the whole system before the summer shutdown.

The objectives of water treatment are the same for these low-pressure boilers as for high-pressure boilers, and so the reader can refer to the treatment details that have already been covered in the section "General Steam Power Plant Operations," which includes the objectives of boiler water treatment. An expenditure for a complete water treatment system consisting of a sodium zeolite softener and equipment for proportional chemical feeding including a water meter, pumps, and tanks is justified for new steel steam or hot water boilers. This is particularly true in the cases of large, more costly boilers of high heat release and extensive loop systems and smaller boilers operating with a high percentage of makeup or using hard-water makeup. In any case, a shot-type feeder (Fig. 4.11) for slugging the chemical solutions directly into the boiler water system should be installed.

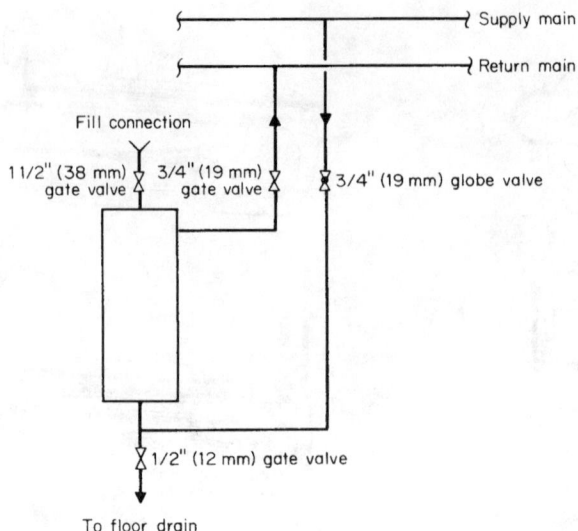

Figure 4.11 Shot-type feeder. (*E. G. Hansen, Hydronic System Design and Operation, McGraw-Hill, New York, 1985.*)

Deaerating feedwater heaters can't usually be justified for the smaller low-pressure systems, but they are a necessity for the larger and more costly systems. Besides the advantage of heating the water entering the boiler and thus reducing boiler strains, these deaerators remove dissolved oxygen down to 0.03 ppm or less and thus reduce the corrosive tendencies of the boiler feedwater.

High-pressure steam boilers[18]

High-pressure steam boilers are defined as boilers operating above 15 psig, but usually they are considered as operating at 100 psig or above. These are usually water-tube boilers, in which steam is formed from water in the tubes by heat input from combustion on the outside of the tubes. The many types of water-tube boilers include the "A" type (Fig. 4.12), which consists of two lower drums and one upper steam drum; the "D" type (Fig. 4.13), a popular package boiler; and the "O" type (Fig. 4.14), which consists of one lower drum and one upper steam drum. They are available in practically any size or pressure desired.

In designing water treatment for a high-pressure steam boiler system, consideration must be given to the system as a whole, the makeup water system, feed lines, the feedwater heater, the economizer, the boiler itself, and the condensate return system. In cogeneration sys-

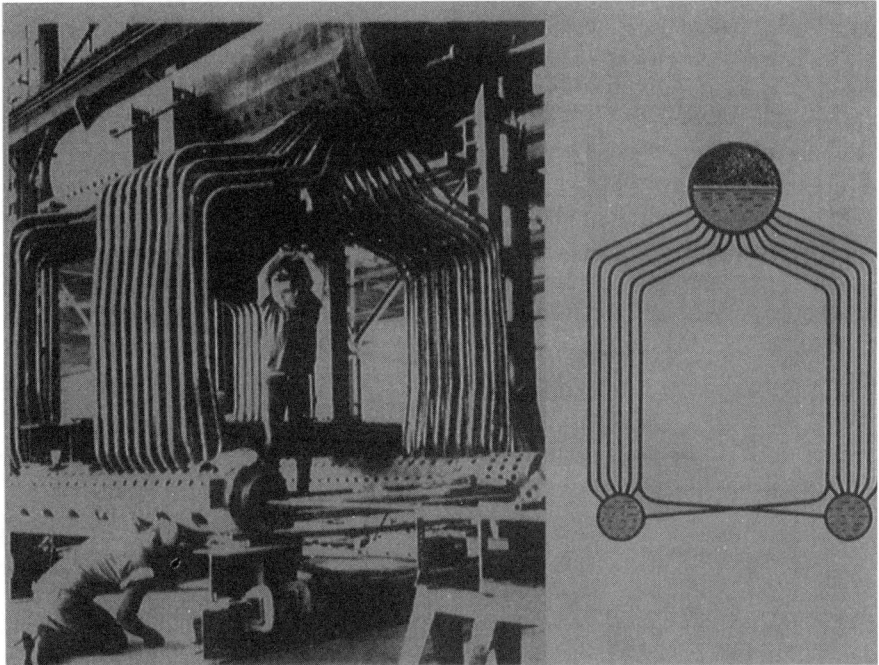

Figure 4.12 Type A water-tube package boiler. [*T. C. Elliott (ed.), Standard Handbook of Powerplant Engineering, McGraw-Hill, New York, 1989.*]

tems, consideration also has to be given to the superheater, reheater, turbine, and turbine condenser.

The influent-water quality, materials of construction in the system, the boiler pressure and temperature, possible contaminants in the pure water system, uses of steam in the system, etc., must all be given attention in designing the external water treatment system and the posttreatment system. The choice of chemicals, points of application, and means and frequency of application must also be given attention.

As a particular installation using a particular water supply may require different choices of materials, chemicals, and points of application, only general advice can be given here. It is suggested that one refer first to Tables 4.3 and 4.6; then decisions can be made as to the extent of sophistication desired in the external treatment system—whether it be reverse osmosis (membranes), demineralization, dealkalization, or softening. The higher the boiler pressure, the more complete the external system and controls should be. Phosphate treatment following zeolite softening is being used very effectively at the lower pressures and even in systems operating above 2000 psig.

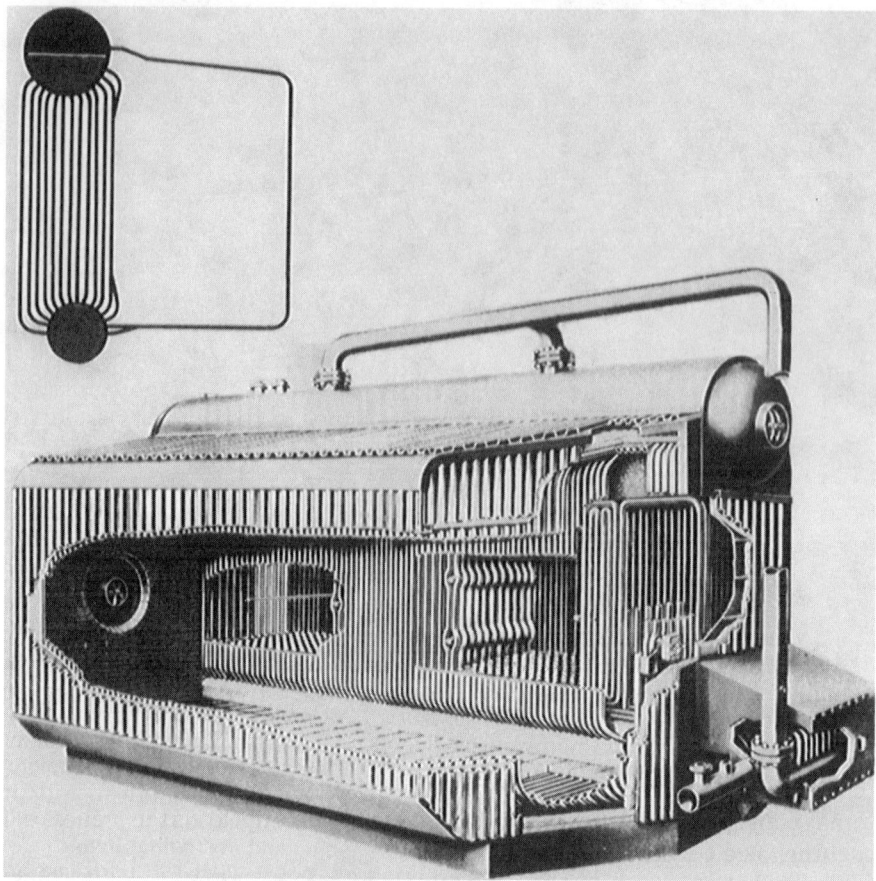

Figure 4.13 Type D water-tube package boiler. [*T. C. Elliott (ed.), Standard Handbook of Powerplant Engineering, McGraw-Hill, New York, 1989.*]

Reverse osmosis is proving to be economical when it comes before demineralization and is even being used in preference to ion exchange at the lower boiler pressures. The steam purity requirements shown in Table 4.8 also need to be considered in deciding on a particular water treatment system.

Electric boilers[19]

In all-electric buildings, electric boilers may be installed and may be the electrode or immersion type. The electrode type uses the water itself as the heating element. This is accomplished by immersion of charged electrodes in the water or by spraying water between oppo-

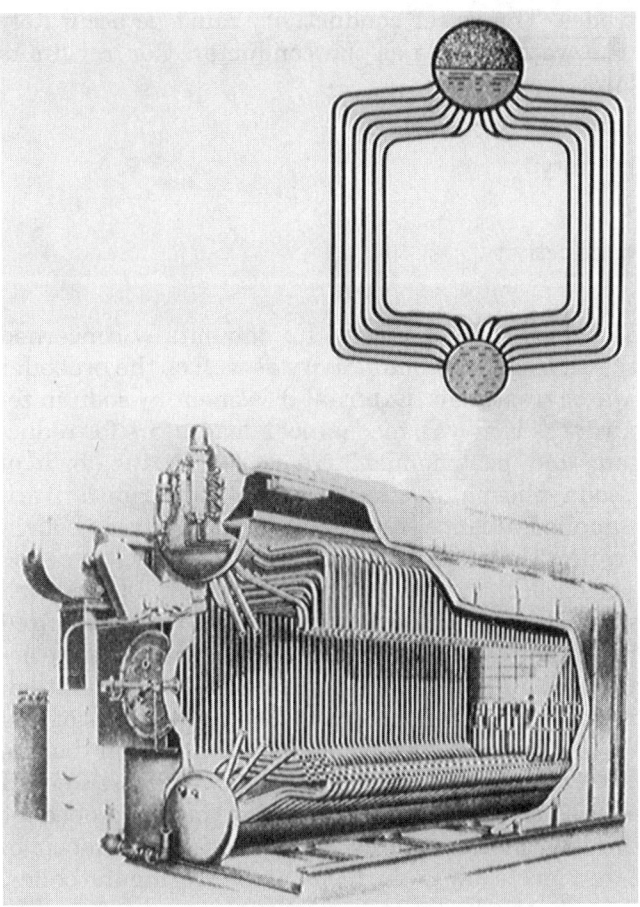

Figure 4.14 Type O water-tube package boiler. [*T. C. Elliott (ed.), Standard Handbook of Powerplant Engineering, McGraw-Hill, New York, 1989.*]

TABLE 4.8 American Boiler Manufacturer's Association Recommendations for Steam Purity

Drum pressure, psig	Total dissolved solids, mg/L		Moisture carryover, %
	Boiler water (max.)	Steam (max.)	
0–300	700–3500	0.2–1.0	0.03
301–450	600–3000	0.2–1.0	0.033
451–600	500–2500	0.2–1.0	0.04
601–750	200–1000	0.1–0.5	0.05
751–900	150–750	0.1–0.5	0.07

sitely charged electrodes. The water conductivity must be accurately controlled because the water serves as the conductor. Control limits normally are as follows:

0.0–3.5 mg/L hardness

200–400 mg/L alkalinity

<5 mg/L iron (Fe)

500–3000 µS/cm conductivity

pH 8.5–10.0

In treating electrode boilers, one should be fundamentally concerned with the boiler water pH, iron, and conductivity as well as the preboiler and internal boiler water treatment. External treatment by sodium zeolite softening for hardness removal, mechanical deaerators for reducing dissolved oxygen, and postchemical treatment in the form of dispersant–caustic soda–phosphate are normally prescribed. Antifoams should not be applied because the circulation in electrode boilers may cause foaming rather than reduce it. Actually, boiler water carryover is not a problem in electrode boilers because steam is produced above the water level, not below the water level as in fossil-fuel-fired boilers. Chelants should not be used because they may cause deterioration of boiler nozzles and other components in the water flow system. Sodium sulfite or hydrazine is used to remove the final traces of dissolved oxygen and to reduce corrosive tendencies. As carbon dioxide from the decomposition of makeup water alkalinity will cause low pH and corrosion of steel in condensate return systems, there is a need for neutralization by the volatile amines (cyclohexylamine and morpholine) to keep soluble iron levels low in the feedwater entering the boiler.

The immersion-type electric boiler is basically a pressure vessel in which a number of resistance-type heating elements are immersed in water. This is the same method as for the residence-sized electric water heaters that have been used for many years throughout the United States. Immersion-type boilers have a relatively high tolerance for varied water conditions, but scaling must be controlled and dissolved solids kept low enough to prevent foaming and carryover. Normally, installation of a sodium zeolite water softener and normal postchemical treatment are adequate to control scaling; however, the possible corrosion of stainless steel by chlorides may make it necessary to install hydrogen exchange softening (sulfuric acid regeneration) instead of salt-regenerated ion exchangers.

The postchemical treatment must be monitored closely in the electrode boilers because of their overall effect on the conductivity. Water treatment methods for the immersion-type boiler are similar to those applied to fossil-fuel-fired low-pressure steam boilers.

Steam Purity and Its Control

The solids content in steam is defined by the purity of the steam. Purity is a measure of the parts per million (ppm) or milligrams per liter (mg/L) of solids impurity in the steam and is determined from precise electrical conductivity, sodium, or chloride measurements. The usual guarantees by the boiler manufacturer call for a maximum of 1 mg/L (1 ppm) dissolved solids when the boiler is operated at a specified load and the feedwater is properly treated within American Society of Mechanical Engineers (ASME) guidelines. The moisture content in steam is defined by its quality, which is a measure of the percentage by weight of dry steam in the mixture. Guarantees by the boiler manufacturer are usually for 99.5 percent steam quality or more, which means that the moisture content is 0.5 percent or less.

Keeping the P alkalinity of the boiler water and conductivity tests and external treatment of makeup water under control is important in ensuring the production of high-quality steam. In addition, hardness and conductivity tests of the condensate and feedwater must be carefully monitored and controlled at minimal levels in order to keep contamination from condenser and heat exchanger leaks at a minimum and makeup water equipment free of malfunctions. Condenser leaks may contribute appreciable water solids contaminants, which can cause increased corrosion, scale, and possible boiler water carryover.

When steam is utilized for the production of electrical energy by a steam turbine, *superheated steam* (steam heated above the saturation temperature, as reported in the steam tables) is normally applied to the steam turbine because of its less erosive characteristics and greater efficiency. Normal boiler steam purity requirements specify that the solids level should be less than 1 mg/L (1 ppm) in order to avoid deposition in the superheater. In addition, desuperheater spray water, used for controlling steam temperature, must be of a quality equal to or better than the required steam purity. The American Boiler Manufacturer's Association (ABMA) lists recommendations for steam purity (see Table 4.8).

Internal baffles and steam separators

Baffles in the boiler drum serve to deflect boiler water from high-heat areas, keeping it from entering the dry pan area. This prevents or reduces boiler water contamination of the high-purity steam leaving the dry pan area. These baffles sometimes develop leaks and allow boiler water droplets to mix with the high-purity steam leaving the boiler. Periodic inspections should be made to determine if baffles are tightly connected to the boiler drum surfaces or if they need repair. Steam separator devices, which operate on the principle of centrifugal force to remove the heavier droplets of boiler water, may also need occasional repair. The operation of feedwater regulators also requires attention to ensure that the boiler water is being maintained at a constant and proper level as specified by the boiler manufacturer. Boiler design should provide for the feedwater to enter the boiler with minimum disturbance of the boiler water level.

Operating variables affecting steam purity[20]

The following conditions can cause boiler water carryover:

1. Operating a boiler above the specified water test limits and at a water level much below or above that recommended by the manufacturer
2. Operating the boiler at loads above the maximum load specified by the boiler manufacturer

Arranging boiler loads so that rapid increases or decreases are minimized reduces the possibility of boiler water carryover. For example, arranging for soot blowing, boiler blowdown, and ash pulling during normal low-load periods reduces the possibility of carryover.

Determination of steam purity

There are a number of analyzers that provide a measure of steam purity by determining conductivity, sodium, and chloride. The one best known is for a conductivity method, as described in ASTM D2186; it is the Larson-Lane Steam Purity Analyzer shown in Fig. 4.15.[20] *Supplement to ASME Performance Test Codes* (Sec. 11: "Water and Steam in the Power Cycle"—concerning purity and quality)[2] describes these in detail. (The author is presently one of the ASME Committee members now updating this publication.) The modern boiler manufacturer seems to be more meticulous in designing boilers free of boiler water carryover problems, and so testing new boilers for steam purity to be certain that they meet the manufacturers' specifications is less necessary now than in the past. However, if turbine or superheater deposit or corrosion problems caused by impure steam arise, such testing can be easily arranged. Probably in most cases, the problem is caused by improper water treatment and control.

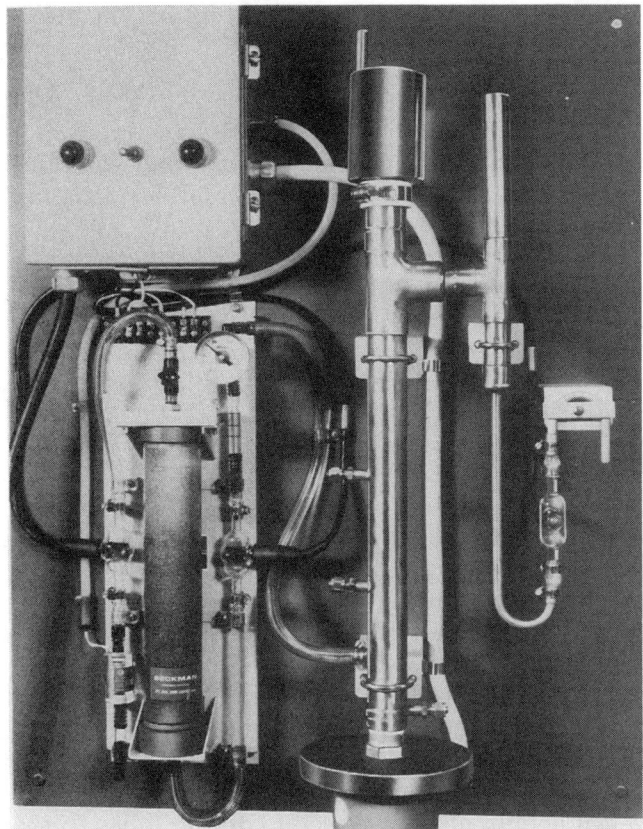

Figure 4.15 Larson-Lane Steam Purity Analyzer, Condensate Reboiler model.

Maintaining Efficient Plant Operation

The following items require attention in keeping the building steam and hot water systems trouble-free:

1. Operation with minimum makeup water usage. Reading the water meter, keeping records of makeup water usage, locating losses, and correcting such losses are important in maintaining efficient plant operation.
2. Return of all uncontaminated steam condensate or loop water.

3. Requirement of properly sized condensate holding or expansion tanks.
4. Proper maintenance and repair of makeup valves and steam traps to avoid leaks and to save valuable condensate.
5. The application of proper water treatment, proper water sampling and water testing, accurate blowdown control, and correction of operations according to the tests.

Operational Techniques for Improving Internal Boiler Conditions

The following techniques should be given attention if internal scale and sludge conditions are unsatisfactory:

1. Consideration should be given to shorter continuous operation and more frequent washouts.
2. If a boiler is removed from service, attention should be directed during the first half hour to applying 5-s blowdowns alternately to top and bottom blowdown valves during six evenly spaced intervals. Then after firing has stopped, blowdown should be continued for longer periods at least every hour so that the boiler water is gradually replaced by fresh feedwater. This procedure should be continued with colder water until the boiler is cool enough to drain; then the boiler should be drained and the drums and tubes flushed with heavy streams of water from high-pressure hoses. Residual sludge should be kept wet; it should not be allowed to dry on tubes. If the boiler is not to be immediately cleaned, it should be refilled with water until it is to be cleaned, but it should never be left filled with water for longer than a few days; otherwise, serious corrosion and pitting will develop.

Boiling Out New Boilers

A new boiler may require an alkaline boil-out before being placed in service. First it should be filled completely (the water level control is blocked off) with hot feedwater and drained. Then after the gage glass is blocked off, the boiler is filled to normal level and is treated with a water solution of 3 lb Item 4 caustic soda, 3 lb Item 7 trisodium phosphate, 1 lb Item 9 sodium nitrate, and 1/4 lb low-sudsing detergent (as Vel) per 1000 gal of boiler water volume. After the stop and check valve is closed, the boiler is fired, and at least 15 psig steam pressure is maintained for 8 to 18 h. "Flash" blowdowns of all blowdown valves should be applied hourly, followed by refilling of the boiler with hot feedwater to the normal boiler water level. Then the fire is shut off, the boiler pressure is reduced by the addition of feedwater, and overflowing is controlled by attaching a temporary line to the top vent of the boiler. Feedwater is applied until the water discharging is "clear"; then the boiler is drained through the bottom blowdown, internal surfaces are flushed with a high-pressure water hose, and internal surfaces are inspected to see that all foreign matter has been boiled out. Since the boil-out solution is strongly alkaline, safety equipment (including a face shield, gloves, and protective clothing) should be used when the water solution is prepared and applied to prevent contact with the skin. Arrangements should be made for alkaline waste disposal according to local environmental discharge requirements.

Proper Lay-up of Boilers When Out of Service[21]

Out-of-service boilers need to be properly stored; otherwise, serious corrosion or pitting will occur. Boilers may be stored wet or dry. Dry storage is the simplest and easiest procedure and therefore is preferred if practical. It is effective only if the internal surfaces of the boiler are kept completely dry. If there is some leakage of stop and check valves, a wet-storage method may be the solution. If the boiler is to be stored without cleaning, then wet-storage method A should be applied. If the boiler is to be stored after cleaning or if it is empty, then method B should be applied. Wet storage is required for boilers with superheaters, since superheater tubing cannot be drained dry. Incorporation of amine treatment (Item 13 cyclohexylamine) is recommended so that distillation of this volatile chemical into the superheater tubing will provide the necessary corrosion inhibition.

Wet storage, method A

The following procedure should be applied for wet storage when the boiler is not to be cleaned internally:

1. Blow down the boiler thoroughly.
2. At least 15 min before the boiler comes off the line, add the following chemicals per 1000 gal of boiler capacity:

   6 lb Item 10 sodium sulfite

   1 lb Item 1F or L236 boiler water dispersants

   Sufficient Item 4 caustic soda (about 8 lb) to provide a P alkalinity reading of 1000–1700 ppm (mg/L)

   Only for boiler with superheater: 1.5 lb Item 13 cyclohexylamine

3. After taking the boiler off the line, fill it to the top, overflowing vents or safety valves to make sure complete filling is accomplished.

4. When the boiler is completely full, close all valves to seal the boiler from the entrance of air. Then test the boiler water every 30 days for sulfite, which should be maintained at a minimum of 50 mg/L. Also test the P alkalinity, which should be maintained at a minimum of 1000 mg/L.

5. Before starting operation, drain the boiler contents to 1/4 drum level, remove the cap from the high-water whistle, and then fill the boiler with feedwater to the regular operating level.

Wet storage, method B

The following procedure should be applied for wet storage after a boiler has been cleaned or if the boiler is empty:

1. Assuming that the boiler has a capacity of 5000 gal, prepare a solution composed of:

   30 lb Item 10 sodium sulfite

   5 lb Item 1F or L236 boiler water dispersants

   40 lb Item 4 caustic soda (sufficient to provide a P alkalinity of 1000–1700 mg/L)

   7.5 lb Item 13 cyclohexylamine (if a superheater is involved).

2. Feed this solution continuously with the feedwater being applied to the boiler.

3. If the boiler has a superheater, start a small wood fire to provide sufficient steam to volatilize the amine over into the superheater. Steam should issue from the superheater vent. After the fire has subsided, it is essential that the boiler be filled completely and that no air pockets be present.

4. During filling, close all valves except those on the top of the boiler. Place a cap on the high-water whistle when water flows out.

5. When the boiler is completely full, close all valves to seal the boiler from the possible entrance of air.

6. During the storage period, keep the boiler full and add chemicals as needed to keep the tests at the specified limits. A small wood fire may be required to provide circulation and uniform distribution of chemicals, particularly if additional chemicals are required.

7. Before starting operation, drain the boiler contents to 1/4 drum level, remove the cap from the high-water whistle, and then fill the boiler with feedwater to the regular operating level.

References

1. E. G. Hansen, *Hydronic System Design and Operation*, McGraw-Hill, New York, 1985.
2. ASME Boiler and Pressure Vessel Committee, *Supplement to ASME Performance Test Codes,* Sec. VI: "Recommended Rules for the Care and Operation of Heating Boilers," Sec. VII: "Recommended Rules for the Care and Operation of Power Boilers," Sec. 11: "Water and Steam in the Power Cycle," American Society of Mechanical Engineers, New York, 1974 and 1989.
3. T. C. Elliott (ed.), *Standard Handbook of Powerplant Engineering,* McGraw-Hill, New York, 1989.
4. *Betz Handbook of Industrial Water Conditioning,* 8th ed., Betz Laboratories, Inc., Trevose, Pa., 1980.
5. *Drew Principles of Industrial Water Treatment,* 1st ed., Drew Chemical Corp., Boonton, N.J., 1977.
6. Nalco Chemical Co., *The NALCO Water Handbook,* 2d ed., McGraw-Hill, New York, 1988.
7. J. Katzel, "Focus on Boilers," *Plant Engineering,* May 24, 1990, p. 68.
8. J. M. Montgomery Consulting Engineers, Inc., *Water Treatment Principles and Design,* Wiley Interscience, New York, 1985.
9. *Consensus on Operating Practices for the Control of Feedwater and Boiler Water Quality in Modern Industrial Boilers,* American Society of Mechanical Engineers, New York, 1990.
10. R. W. Lane, *Industrial Water Treatment Guidelines and Water Analytical Methods,* Champaign, Ill., 1983.
11. *Pamphlets on Water Treatment Submitted to U.S. Air Force on USAF Contract FO8635-79-CO161,* Illinois State Water Survey, Champaign, Ill., 1979.
12. J. N. Tanis, *Procedures of Industrial Water Treatment,* Ltan Co., Ridgefield, Conn., 1987.
13. R. Holzhauer, "Hot Water Boilers," *Plant Engineering,* May 28, 1987, p. 34.
14. S. A. Edgerton, D. V. Kenny, and D. W. Joseph, "Determination of Amines in Indoor Air from Steam Humidification," *Environmental Science Technology,* vol. 23, no. 4, 1989, p. 484.
15. J. J. Halas, "Reflections on Steam-Humidified Room Air," *Engineered Systems,* Parts 1 and 2, February 1991, p. 98, and March 1991, p. 72.
16. S. Strauss, "Concern Rises for Safety in Feedwater Deaerators," *Power,* November 1983.
17. *1991 Annual Book of ASTM Standards,* vol. 11.01: *Water* (I), American Society for Testing and Materials, Philadelphia, p. 161.
18. R. D. Port and H. M. Herro, *The Nalco Guide to Boiler Failure Analysis,* McGraw-Hill, New York, 1991.
19. T. J. Heil and C. H. Leatham, Jr., "Electric Boilers up to 175000 lb/hr," American Power Conference, Chicago, Apr. 19, 1977.
20. R. W. Lane and G. Otten, *Power Plant Instrumentation for Measurement of High-Purity Water Quality,* STP 742, American Society for Testing and Materials, Philadelphia, 1981.
21. *Boiler Layup Procedures,* American Society of Mechanical Engineers, New York, 1983.

Chapter

5

Open Recirculating Cooling Water Systems and Treatment

Open recirculating cooling water systems in the form of cooling towers, evaporative coolers, or evaporative condensers are water-saving devices for providing cooled water or cooled refrigerants for air conditioning or refrigeration systems. Instead of passing once through these systems to waste, the water for the chillers runs through cooling towers, where 8 to 15°F (4.5 to 8.3°C) cooling is provided before return use. This provides a savings of 95 percent of the water circulated. Essentially, cooling towers are simple means for obtaining cooled water by evaporation.

Originally, the term *a ton of refrigeration* referred to the energy per day required to freeze one ton of water into ice at 32°F (0°C); however, the cooling tower industry now defines it as 15,000 Btu/h. Cooling chiller refrigerants with air has become a more common practice for the smaller comfort-cooling installations (<300 tons), and while this is a less efficient means of cooling, the advantages are that water treatment and attention to the rigid necessary control of water treatment are not required.

The use of once-through cooling water is practiced in some cases where plentiful and inexpensive makeup water supplies are available (for example, the Chicago River is used for this purpose by many of the Chicago Loop skyscrapers). Biocide treatment to prevent the formation of an insulating biofilm on heat transfer surfaces is still required, and strict attention must be directed toward keeping treatment levels within U.S. Environmental Protection Agency (EPA) regulations. Chemical costs for treating once-through systems can be appreciable owing to the large volumes of water involved. In the case of a corrosive water, an alternative and more economic solution may be the installation of more expensive corrosion-resistant materials for eliminating the possible occurrence of serious corrosion. In the case of waters of high hardness and alkalinity, or brackish water and seawater, the application of scale- and corrosion-inhibiting chemicals may likely be required, even in the once-through systems.

The installation of cooling towers using water rather than air as the cooling medium is usually preferred because of the lower cost of operation. It is reported that there are now 500,000 cooling towers in the United States.[1] They are designed so that countercurrent airflow causes evaporation and cooling as the water passes through the tower and is circulated back to process cooling. Cooling towers are designed so that maximum evaporative surface is provided for the water in passage down through the orifices in the distribution deck, over the wood or plastic slats (called *fill*), and into the storage basin. In passage through the towers, 1.2 percent of the flowing water is evaporated for each 10°F (5.6°C) temperature decrease. Actually, 75 percent of the cooling results from release of the latent heat of vaporization, and 25 percent results from sensible heat transfer.

The decrease in temperature as the water passes through the tower is called the *range,* and the difference between the effluent temperature and the wet-bulb temperature is called the *approach* to the wet bulb. Specification details for the purchase of cooling towers reveal that a finite quantity of water is to be cooled from a particular temperature to a definite lower temperature at a specified wet-bulb temperature.

Free, or Ambient, Cooling

Microprocessor-based free cooling systems have been installed in a number of large air-conditioning systems in areas of the United States where the outside temperatures in the colder season can lower cooling tower water temperatures adequately. Essentially, the term *free cooling*[2] means that cooling tower water is added directly to the closed system in which chilled water from the chillers is normally circulated. An alternative is to circulate the cooling tower water through heat exchangers (rather than through the chillers) for cooling the closed-system chilled water indirectly. During in-between seasons (spring and fall), when somewhat higher chilled-water temperatures can be maintained and when there is little need for dehumidification, the cooling tower water may be cool enough to eliminate the need for operation of the chillers and compressors. Appreciable energy savings may be effected by the use of direct cooling or through the use of plate or other heat exchangers (see Figs. 6.1 and 6.2).[2] For example, automatic control valves may channel the cooling tower water into the building's chilled-water loop or a heat exchanger system when a manually set [for example, 52°F (11°C)] wet-bulb air temperature is registered. A defrost feature may also be included in the system to prevent cooling tower water from freezing in the winter.

Since the water treatment requirements for an open recirculating system (cooling tower water) are different from those for a closed recirculating system, the consequences of adding an aerated water supply containing suspended matter to a closed water system designed for the use of minimal noncorrosive makeup water must be given extra attention. The installation of a filter (perhaps a sidestream filter, filtering 5 percent of the flow) for the cooling tower water may be *particularly* justified if free cooling is planned. In fact, the installation of a combination of strainers and filters[3] may be justified for many cooling tower systems if there is evidence of appreciable suspended solids in the circulating water. Strainers remove particles down to 45 µm but serve the purpose of removing the larger particles so that filters can more effectively remove particles down to 5 µm. Self-cleaning strainers and filters are also available; these relieve the maintenance problem of replacing cartridges or backwashing. In order to keep the strainer or filter sizes and costs more reasonable, it can be decided to install this equipment in a sidestream position so that only 3 to 7 percent of the flow of circulating water is filtered while a very beneficial effect is still provided.

Strict adherence is required in maintaining the levels of the cooling water treatment near the upper limits prescribed for treatment control of cooling tower water. Increased water treatment inhibitors may be required if corrosion probe or coupon results indicate that increased corrosion is occurring, if hardness tests indicate that scale inhibitor treatments are inadequate, or if increased bleedoff is necessary. This will be discussed later in this chapter in the section "Water Treatment and Water-Testing Equipment."

Cooling Towers

The two common forms of cooling towers are counterflow and crossflow, as illustrated in Figs. 5.1 and 5.2.[1] As vapor from the evaporation of the water leaves the water surface, the concentration of solids (including hardness) in the circulating water builds up. This must be taken into consideration in designing water treatment for scale and corrosion inhibition and particularly in designing bleedoff control. Controlled bleedoff to maintain at least three to four cycles of concentration (3 to

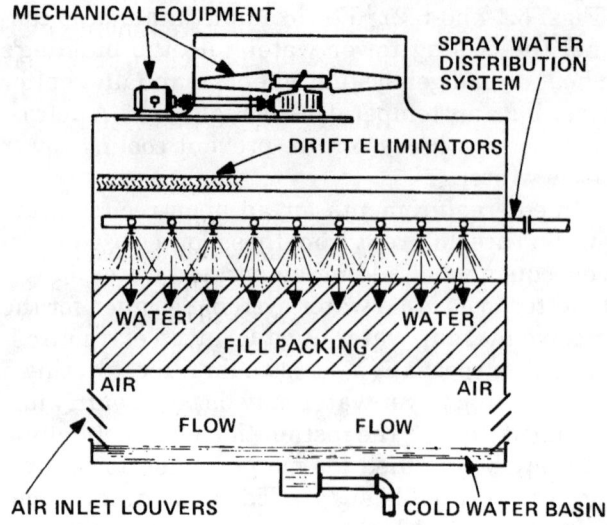

Figure 5.1 Counterflow cooling tower. (*R. Burger, Cooling Water Technology, Dallas, Tex., 1979.*)

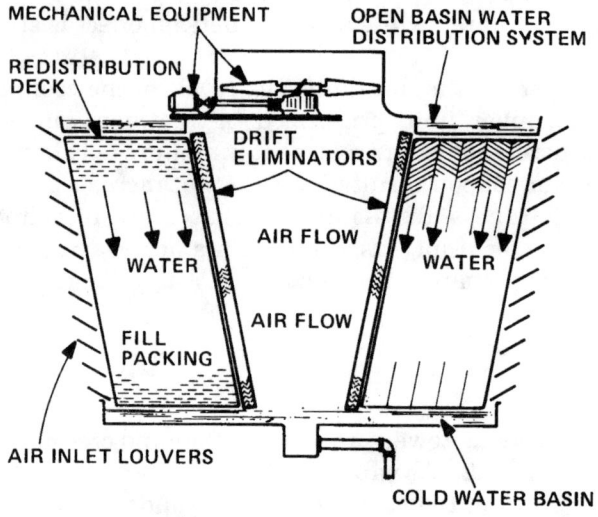

Figure 5.2 Crossflow cooling tower. (*R. Burger, Cooling Water Technology, Dallas, Tex., 1979.*)

4 times the original water hardness, chloride, or conductivity) is normally prescribed, along with adequate chemical water treatment for attaining effective control of corrosion and scale and efficient operation. Makeup water is added to replace the water lost by evaporation, by drift of moisture from the top and sides of the tower, by windage, and by bleedoff. The quality of the makeup water with respect to hardness (calcium), pH, alkalinity, and dissolved solids (conductivity); the efficiency of the treatment; the cleanliness of the air; and the suspended matter in the circulating water all affect the allowable cycles of concentration. In addition, carbon dioxide in the circulating water is lost in passage through the cooling tower; as a result, a pH increase to 8.3 to 9.0 occurs, tending to make the water more scale-forming and more difficult to treat. Just remember that all water is corrosive and that the metal (not the water) is actually being treated to provide a protective oxide film.

The location of cooling towers may influence their service and corrosion resistance. Other nearby industries' discharge and the discharge from one's own installation (for example, kitchen plumes, laboratory acid fumes, and fossil plant flue gas) may load the circulating air with vapors, which are acidic, alkaline, grease-ridden, and high in carbon and which contain suspended matter or other nutrients supporting bacterial growth in the cooling tower water. Such vapors may make it necessary to neutralize acidic or alkaline conditions in the circulating water, apply a different or more complex biocide treatment program, and install in-line filters to remove corrosive material and suspended matter.

A cooling tower that is in continual contact with flowing corrosive water, which is saturated with dissolved oxygen, is in a particularly corrosive environment. There may not be appreciable steel or galvanized-steel surfaces in a cooling tower except for the structural members and the piping; however, for long-life service, steel should actually have a protective coating like a rubber-based polymer. It should also be recognized that such coatings must be absolutely complete; otherwise, corrosion at holidays in the coating may develop into serious pits.[1]

Galvanized-steel towers

Galvanized-steel towers are getting a lot of attention now because the deposition of so-called white rust (predominantly basic zinc carbonate) has been observed on their surfaces. One reason for this may be that some of the galvanized steel, which is being imported, is not of as high a quality as it used to be. Also, in the past the galvanized steel in the older towers received a final chromate rinse after being dipped in molten zinc. A higher pH is now being maintained in cooling towers because the U.S. Environmental Protection Agency (EPA) has banned the use of zinc-chromate treatments in comfort-cooling towers, and the presently used corrosion inhibitors must be used at a higher pH to be effective. Under higher pH conditions it would be expected that zinc would be more soluble and would more readily form the basic zinc carbonate. The solution[4] is to operate the towers that have developed "white rust" at a somewhat lower pH of 7.8 for about four months, using phosphate or molybdate treatment until the "white rust" gradually disappears. After this, the pH can be increased to 8.5 without the development of "white rust."

One method that has also been proposed for starting treatment in a new galvanized-steel tower is to circulate a solution composed of 2 lb Item 31 sodium acid phosphate (NaH_2PO_4) and 0.1 lb Item 241 surfactant per 1000 gal through the tower system for several hours. This causes a film of zinc phosphate to develop on the surface of the galvanizing and prevents the later formation of basic zinc carbonate. Subsequent treatment should consist of conventional treatment containing a molybdate or orthophosphate corrosion inhibitor plus phosphonate, an acrylate polymer for scale control, and azole for corrosion inhibition for copper-bearing metal. Also, the treatment should be designed for operation at a maximum pH of 8.5.

Cleaning of a galvanized-steel tower which had become coated with "white rust" has also been observed to be effectively accomplished by the application of a smooth black protective coating.*

Materials for cooling towers

The structural materials in common use are galvanized steel, stainless steel, fiberglass (a matrix of fiberglass-reinforced polyester), and wood. Galvanized towers with stainless-steel basins are fairly common, but it should be pointed out that stainless-steel basins may be subject to pitting if they are allowed to collect debris on their surfaces without periodic cleaning. In general, stainless-steel surfaces need to be kept clean so that oxygen can contact the surfaces and preserve the protective oxide surfaces. Allowing crud to be retained on practically all metallic surfaces may lead to the development of differential aeration cells and pitting corrosion.

It is typical to have a fiberglass exterior casing for some wood and galvanized towers. For towers up to 1000-ton cells, galvanized steel is the least expensive material, particularly because it lends itself to factory assembling. Higher in cost is fiberglass and then stainless steel.

---

*The trade name is Belzona. This underwater formulation is manufactured by Rumford Industrial Group, Inc., P.O. Box 670, 772 Congress Park Drive, Centerville, OH 45459.

Wood becomes the least expensive material in towers consisting of 1000-ton cells or above.[1]

Hardware selections are usually the same as the choice of structural material. Properly coated steel is the least expensive and should be durable. The 300 and 400 series stainless steels are generally more corrosion-resistant and more expensive, and 316 stainless steel is even more corrosion-resistant and expensive.

Current fill made from properly formulated PVC sheets proves to be a durable choice. Water distribution nozzles made from polypropylene, ABS (acrylonitrile-butadiene-styrene), or fiberglass-filled nylon have been shown to possess proper chemical and erosion resistance.

It may prove prudent to let the cooling tower manufacturer specify the particular materials to be included in a cooling tower. The manufacturer is probably best qualified to choose materials that will be the most economical and efficient for a particular environment. Of course, review of the manufacturer's specifications by a qualified materials specialist is also in order.

Calculation of cooling tower treatment and blowdown[5–9]

It is necessary to know the amount of evaporation (E), the amount of makeup (M), and the amount of blowdown (B) in addition to the water quality to best design a water treatment program for a cooling tower. These values may be calculated as follows:

Evaporation (E) in absorption air-conditioning systems operating at full load is approximately 3 gal/(h)(ton); in centrifugal or reciprocating systems, E is approximately 1.5 gal/(h)(ton).

Makeup water (M) = evaporation (E) + blowdown (B) (includes drift, which is water carried away by fans and wind).

Cycle of concentration (COC) is the number of times that the original water solids concentrate as water evaporates in the cooling tower. It is calculated according to the following formula:

$$COC = H_{ct} \text{ or } CO_{ct} \text{ divided by } H_M \text{ or } CO_M$$

where H_{ct} and CO_{ct} are the hardness and conductivity of the cooling tower water and H_M and CO_M are the hardness and conductivity of the makeup water.

$$M \text{ (makeup)} = E + B = E + COC/(COC - 1)$$

$$B \text{ (blowdown)} = E/(COC - 1) \text{ or } M/COC$$

Example An absorption air-conditioning system of 200 tons.

$$E = 200 \times 3 = 600 \times 24 \text{ h} = 14{,}400 \text{ gal/day}$$

If the system is operated at 4 cycles of concentration, blowdown is

$$B = 14{,}400/(4 - 1) = 4800 \text{ gal/day}$$

Then

$$M = 14{,}400 + 4800 = 19{,}200 \text{ gal/day to be treated}$$

Tests are conducted to determine the stability of the circulating water in relation to scale formation. For example, if the hardness per cycle of concentration is lower than the hardness per cycle of the makeup water, it may be concluded that scale or hardness sludge formation is occurring in the tower system.

Example

Makeup water (M) hardness (H): 120 mg/L (ppm) (as $CaCO_3$)

M conductivity: 300 μS/cm

H/COC = 120/300 = 0.4 for 1 COC

Cooling tower (CT) hardness (H): 500 mg/L (as $CaCO_3$)

CT conductivity: 1500 μS/cm

H/COC = 500/1500 = 0.3; should be 0.4, indicating that hardness is lower than it should be for 1500/300 or 5 COC, and so hardness has precipitated as sludge or is contributing to scale formation on heat exchanger surfaces.

Conclusion: Scale inhibition is inadequate. Perhaps scale inhibition can be improved by inhibitor feed increase or substitution of a more effective inhibitor, increased bleedoff, or initiation of acid treatment.

An alternative method of determining the number of cycles of concentration is to subtract the conductivity of the added treatment from the makeup conductivity and the circulating-water conductivity according to the following equation:

$$\text{COC} = [e - (\text{COC} \times g)]/(f - g)$$

where e = cooling-water conductivity = 2000 μS/cm

f = makeup conductivity = 500 μS/cm

g = conductivity of added treatment = 25 μS/cm

For example, for a COC of 4:

$$\text{COC} = (2000 - 100)/(500 - 25) = 4.0$$

Water Treatment and Water-Testing Equipment[10–14]

One of the purposes of this book is to impress architects and building designers with the importance of installing proper and adequate water treatment and water-testing equipment. With such equipment, good maintenance of the air-conditioning equipment can more easily be provided. *Conscientious attention to the proper application and control of the cooling water treatment is essential to prevent serious scale and corrosion problems and to prevent excessive maintenance costs.*

Reliable automatic proportioning and pumping equipment actuated by the makeup water meter should be installed for the proper application of the chemical treatment. The required equipment should include an electricontact water meter that initiates chemical pump feeding whenever a set amount of gallons (perhaps 50 to 500 gal) of makeup water is used or every few minutes as specified. Modern proportioning equipment is automatic and provides continuous and exactly proportionate feed on the basis of makeup water usage. This equipment also should include automatic blowdown (bleedoff) control that is preferably actuated by makeup water usage rather than by conductivity or that is actuated following chemical pump application periods or by a timer setting.

Basing the chemical feed proportioning on the air-conditioning load (as indicated from makeup water usage according to the water meter) is the best way to control chemical feed and bleedoff. Auxiliary equipment such as timers, corrosion-resistant chemical pumps, tanks, mixers, and solenoid blowdown valves must be included so that accurate chemical treatment and blowdown control can be provided (see Figs. 5.3 and 5.4). Modern proportional feeding equipment provides digital makeup water meter and blowdown volume readings. The chemical

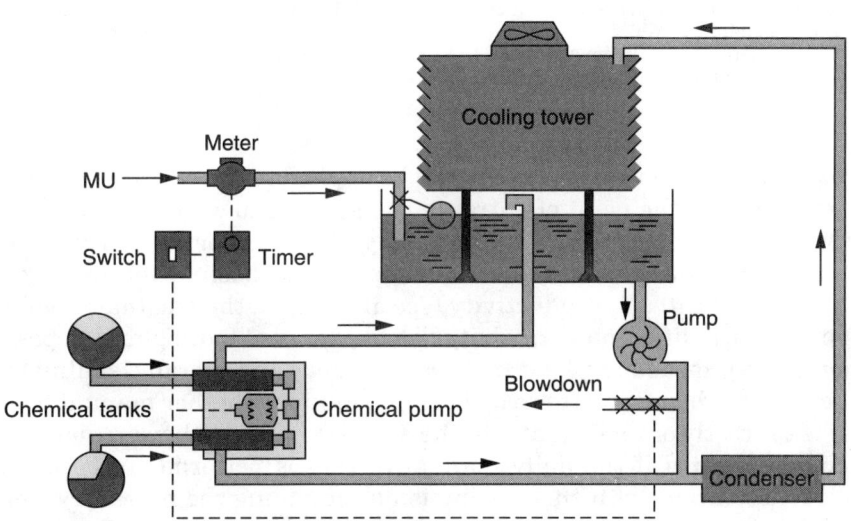

Figure 5.3 Meter-controlled cooling tower treatment. (*R. W. Lane files.*)

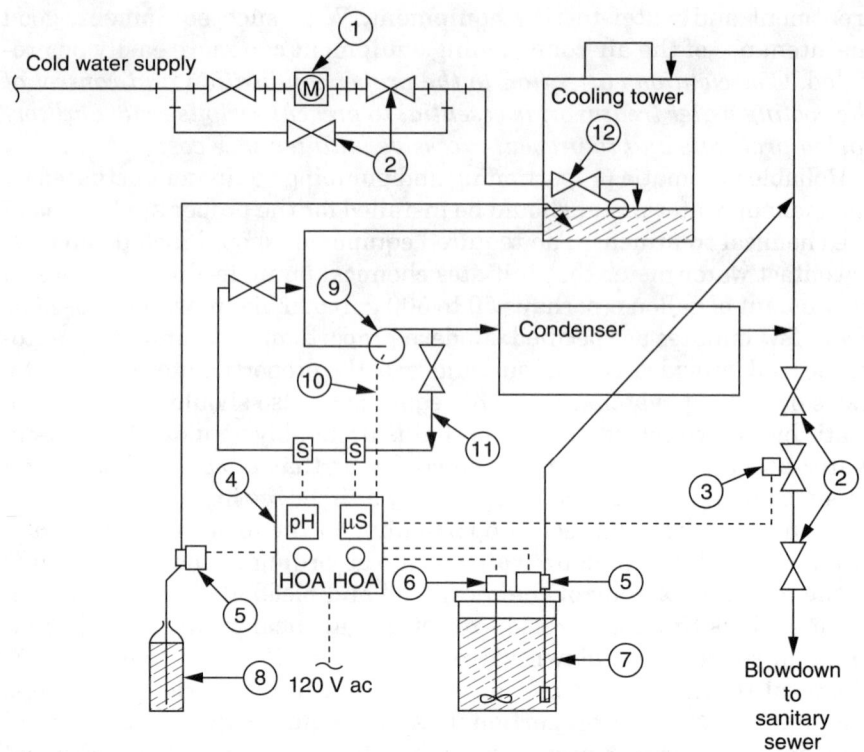

Figure 5.4 More complex cooling tower treatment system: conductivity-pH treatment control. (1) cold water meter; (2) gate valve; (3) electrically operated solenoid valve; (4) treatment controller, pH and conductivity; (5) chemical metering pumps; (6) chemical mixer; (7) chemical inhibitor vat; (8) acid carboy; (9) recirculating cooling water pump; (10) treatment system interlock with 9; (11) sensor recirculating piping; (12) cooling tower makeup float valve. (*Illinois State Water Survey, Champaign, Ill.*)

treatment solution should preferably be applied to the basin of the cooling tower near the point of entrance of the makeup water.

Modern chemical feed systems[13] may obtain signals from various sensors, such as those responding to time, flow, basin level, temperature differential, pH, conductivity, specific ions in the treatment being applied, turbidity, and corrosivity. There are also fouling and deposit monitors for measuring heat transfer efficiency and overall cleanliness. The need for installing these devices depends on whether manual testing is too costly or inadequate. In the author's opinion, they are not particularly needed if the daily manual testing is performed accurately and conscientiously. If this testing is not providing the necessary control to keep the cooling system free of corrosion, scaling, or fouling problems, then installation of a proper monitor may be in order. In the case of a serious biological problem, which may involve microbe-induced corrosion (MIC) and sulfate-reducing bacteria (SRB), installation of a biological monitor or fouling monitor[13] should be given attention. It is possible to rent such monitors (including interpretation and diagnosis) from the various water treatment companies if the need is not expected to be permanent. The need for the more exotic treatment equipment will depend on:

1. The water quality and variation in quality
2. The particular cooling process involved and how precise the control must be
3. The temperatures required in the system
4. The type of treatment to be used in the system, particularly if it is an acid-type treatment

Usually the systems shown in Figs. 5.3 nd 5.4 are adequate, although pH control may also be necessary if the water quality variables and process demands reveal the need for more exact control. High-hardness waters [above 200 mg/L hardness (as $CaCO_3$)] may require acid treatment to properly control scale formation, and in this case, pH control equipment is a necessity. If the cooling water is operated above the common temperatures of 90 to 110°F (32 to 38°C) required in comfort cooling, installation of more complex water treatment control equipment and chemicals will likely be required.

Powdered or more concentrated chemical treatments may be purchased by specification of active ingredients. Reference to App. A, which lists the detailed specifications of various chemical treatments, will aid the user in purchasing these items. The advantages of purchasing and applying generic chemicals according to such specifications are as follows:

1. Reduced cost
2. Attainment of a better understanding of the exact role of specific chemicals in controlling scale, corrosion, and fouling, which may occur in a cooling water system
3. Better control through the adjustment of specific individual chemicals required to correct a problem rather than through adjustment of the treatment level of the overall mix

The installation of chemical vats and electric mixers is necessary to ensure the complete dissolution of separate generic chemicals, which are less expensive when in bulk.

Objectives of Cooling Water Treatment[8,9,12]

The objectives of cooling water treatment are as follows:

1. Prevention and control of scale formation
2. Prevention and control of corrosion
3. Prevention and control of fouling caused mainly by algae and slime deposits

Prevention and control of scale formation

There are a number of methods of pretreatment (or external treatment) for preventing and controlling scale formation. For large installations using untreated turbid water sources, the need for pretreatment methods of aeration, prechlorination, sedimentation, and clarification should be considered. Turbid river or lake waters in particular may require one or more of these pretreatment methods.

Cold lime–soda softening has the advantage of reducing the hardness and alkalinity of a high-hardness water supply and providing a more acceptable quality of water. This will reduce the costs of internal water treatment and provide more economical levels of bleedoff. The following equations show the reactions of makeup water hardness with lime and soda ash in cold lime–soda softening:

$$Ca(HCO_3)_2 + Ca(OH)_2 \rightarrow 2CaCO_3 + 2H_2O$$

Calcium bicarbonate + hydrated lime → calcium carbonate + water

$$CaSO_4 + Na_2CO_3 \rightarrow CaCO_3 + Na_2SO_4$$

Calcium sulfate + soda ash → calcium carbonate + sodium sulfate

Since calcium carbonate is insoluble, it is allowed to precipitate and settle out or is removed by filtration. Since lime soda–softened water often tends to continue to soften or precipitate calcium carbonate (form calcium carbonate scale), recarbonation from the addition of carbon dioxide, acid, or polyphosphate may be required as aftertreatment to reduce the pH and hardness-precipitating tendency. It may likely be necessary to determine the Langelier Saturation Index and to refer to the computer programs in App. D to decide on the need or extent of need for supplemental chemicals or treatment.

An alternative method of softening is to pass the hard water through a sodium ion exchange softener (refer to Fig. 4.3), as shown in the following equation:

$$Ca(HCO_3)_2 + Na_2Z \rightarrow CaZ + 2NaHCO_3$$

Calcium bicarbonate + sodium zeolite → calcium zeolite
+ sodium bicarbonate

Regeneration with salt makes the ion exchanger ready for use again:

$$CaZ + 2NaCl \rightarrow Na_2Z + CaCl_2$$

Calcium zeolite + salt → sodium zeolite + calcium chloride

A less common ion exchange technique is to pass the water through a dealkalizer, which reduces the alkalinity and scaling tendency of the water.[12]

A common method of reducing calcium carbonate scaling tendency is to add sufficient acid (usually sulfuric acid, which is inexpensive). This partially reduces the total alkalinity and scaling tendency of the makeup water, as illustrated in the following equation:

$$Ca(HCO_3)_2 + H_2SO_4 \rightarrow CaSO_4 + 2H_2CO_3$$

Calcium bicarbonate + sulfuric acid → calcium sulfate + carbonic acid

As shown in the paragraphs below, adding acid to reduce the M alkalinity from 100 mg/L to 20 mg/L increases the allowable cycles from 3.3 to 7.0 without calcium carbonate scaling occurring. The application of acid treatment is considered somewhat unsafe; handling sulfuric acid necessitates the use of protective clothing, such as goggles, rubber gloves, and an apron, to prevent the corrosive effect of the acid on skin and clothes if spilling or leakage occurs. Also, close monitoring and control of the application of acid is necessary to prevent the corrosive effect of the acid on exposed metals in the system. Presently, sodium zeolite softening for treatment of hard waters is being given more consideration as a substitute for acid treatment. While close or frequent monitoring is not required, the necessary monitoring may still be difficult to arrange.

Scaling tendency of calcium carbonate and calcium sulfate. Calcium carbonate and calcium sulfate are the two most likely deposits to form in cooling tower systems. There are a number of indexes that indicate the scaling tendency of calcium carbonate and calcium sulfate in water. These indexes prove useful in the design of cooling water treatment for scale inhibition.

The *Langelier Saturation Index (LSI)*,[15] which was described in detail in Chap. 3 and was the first recognized index, is calculated from knowledge of the calcium, alkalinity, dissolved solids, and temperature of the environment. It is based on the solubility of calcium carbonate at a known temperature and dissolved solids content and is calculated through determination of the pH of saturation pH_s of calcium carbonate:[15]

$$pH_s = A + B - [\log \text{Ca (as CaCO}_3) + \log \text{M alkalinity (as CaCO}_3)]$$

A is obtained from Table 3.2 and is estimated from the temperature of the water; B is obtained from Table 3.2 and is estimated from the dissolved solids content (or conductivity).

Example A water contains 100 mg/L calcium (as $CaCO_3$) (calcium is usually about 60 to 65 percent of the hardness content), 100 mg/L M alkalinity (as $CaCO_3$), and 400 mg/L dissolved solids, calculated from conductivity (615 µS/cm × a factor of 0.65 = estimated 400 mg/L dissolved solids). At a temperature of 90°F (32°C), this water would have a pH of saturation pH_s of 1.85 + 9.86 − (2 + 2), or 7.7; then, subtracting this value from the actual pH (8.0) at the site:

$$LSI = pH - pH_s = 8.0 - 7.7 = +0.3$$

The +0.3 indicates a slight scaling tendency in the absence of scale inhibitors; positive values indicate scaling tendency, while negative values indicate corrosive tendency.

The *Ryznar Index* (*RI*), described in Chap. 3, is calculated from the equation $(2 \times pH_s) - pH$, and so the RI for the above water would be 7.4:

$$(2 \times 7.7) - 8.0 = 7.4$$

Values below 6.0 indicate a scaling tendency, and values above 6.0 indicate a tendency to dissolve calcium carbonate.

The *Puckorius Index* (*PSI*),[16] based mainly on practical observations of acid-treated cooling tower waters, is useful in providing a more accurate estimate of the scaling and corrosive tendencies of makeup waters. It is calculated as follows:

$$PSI = (2 \times pH_s) - [1.465 \times \log \text{M alkalinity (as CaCO}_3)] - 4.54$$

The PSI of the above water = 15.4 − 2.93 − 4.54 = 7.97, indicating a tendency to dissolve calcium carbonate. PSI values of 6.0 down to 3.0 indicate increased scaling tendency, while values of 6.0 to 10.0 indicate a stronger scale-dissolving tendency and corrosive tendency.

It is not unusual to find that these indexes may yield contradictory results, but practical observations have revealed that the Ryznar and Puckorius values are more indicative of actual events.

Item 95 (60% HEDP) is the most common phosphonate used for inhibiting scale formation by calcium carbonate. The following data[16] show the scale-inhibiting ability of Item 95 in relation to the Ryznar and Puckorius indexes:

mg/L Item 95 in water at room temperature	RI or PSI value below which scale will form
0.0	6.0
1.0	5.5
2.0	5.1
3.0	4.6
5.0	4.0
10.0	3.9

Additional information on the various indexes is provided in Chap. 3.

The phosphonates [Item 84 sodium aminotrimethylphosphonate (AMP) and Item 95 diphosphonic acid (HEDP)], the acrylates (Item L218A), and the polyphosphates (Items 8 and 58) are effective inhibitors of calcium carbonate scale formation at a dosage of 2 to 6 mg/L (0.02 to 0.05 lb/1000 gal). The LSI, RI, and PSI values obtained in the presence of these chemicals do not provide an accurate indication of calcium carbonate scaling tendency. However, *the following equation* does provide a good estimation of the scaling tendency and the maximum cycles of concentration (COC) at which a phosphonate-treated cooling water can be maintained without serious calcium carbonate scale formation occurring:

Maximum COC = the square root of [110,000/(Ca) × (M alkalinity)]

Example If a water contains 100 mg/L calcium hardness and 100 mg/L M alkalinity (as $CaCO_3$), then COC = 3.3, showing that this water should be blown down to keep the cycles of concentration below 3.3 or, for example, to keep the conductivity of the tower water at a maximum of 3.3 times the makeup water conductivity.

The above example shows that this type of treatment is limited to waters of low to medium hardness, as operating below 3 concentrations is considered uneconomical. If sulfuric acid is added to lower the total M alkalinity of this water, operation at a higher cycles of concentration would be possible without scale formation. However, it might be decided that operating at a lower cycles of concentration is preferable to handling sulfuric acid, which can be considered somewhat hazardous. If it was decided to add acid to lower the total M alkalinity from 100 mg/L to 20 mg/L, then *theoretically, the cycles of concentration could be increased to 7 (the square root of 110,000/2000).*

While applying sulfuric acid to reduce calcium carbonate scale formation is extremely effective, strict safety precautions are necessary in handling sulfuric acid. However, one of the advantages of acid treatment is that corrections for undertreatment can be made by brief application of a slight overtreatment of acid. This practice of overfeed of acid should generally be avoided, as it can cause serious corrosion.

Calcium carbonate scale control.[15] The common scale-inhibiting chemicals used for internal treatment of cooling water systems are as follows (Fig. 5.5):

Phosphonates, such as Item 84 AMP (pentasodium aminotrimethylphosphonate) and Item 95 HEDP (diphosphonic acid)

Polyacrylate copolymers, such as Item L218A Rohm & Haas Acumer 3100 and Item L214 Goodrite K-796

Controlled sulfuric acid, added to reduce pH and M alkalinity

Slightly soluble sodium polyphosphate (Item 58), for use in smaller towers and simpler water treatment

The phosphonates do degrade to orthophosphate in time and also are degraded by ultraviolet light, but the polyacrylates do not. Complete specifications for the above chemicals are given in App. A.

Zeolite softening (sodium ion exchange; see Fig. 4.3) of makeup water[14] is an effective means for inhibiting scaling tendencies and has the advantage of not requiring acid use, which can be hazardous. It is a somewhat more expensive method that requires effective corrosion

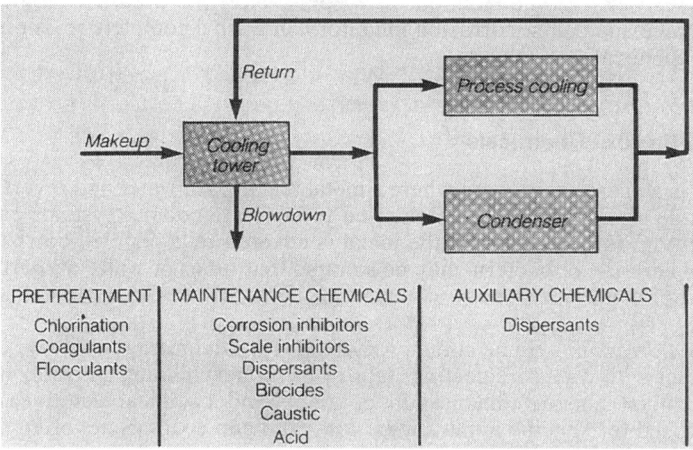

Figure 5.5 Key cooling water chemicals. [*T.C. Elliott (ed.), Standard Handbook of Powerplant Engineering, McGraw-Hill, New York, 1989.*]

inhibition (see the section "Prevention and Control of Corrosion" in this chapter); also, it may require somewhat increased bleedoff and may limit cycles of concentration because of dust (dirt) accumulations. It may cause some dissolving of calcium carbonate from concrete basins and delignification of cooling tower wood owing to the high pH and alkalinity resulting from the softened makeup water. Softening to about 30 mg/L hardness content (as $CaCO_3$) or resorting to complete softening only 90 percent of the time rather than continuous complete softening (to 0.0 mg/L) is recommended to reduce corrosive tendencies.

Calcium sulfate scale control. As calcium sulfate is much more soluble than calcium carbonate (Fig. 5.6), (2000 mg/L compared with 35 mg/L for calcium carbonate), controlling calcium sulfate scale is not as diffi-

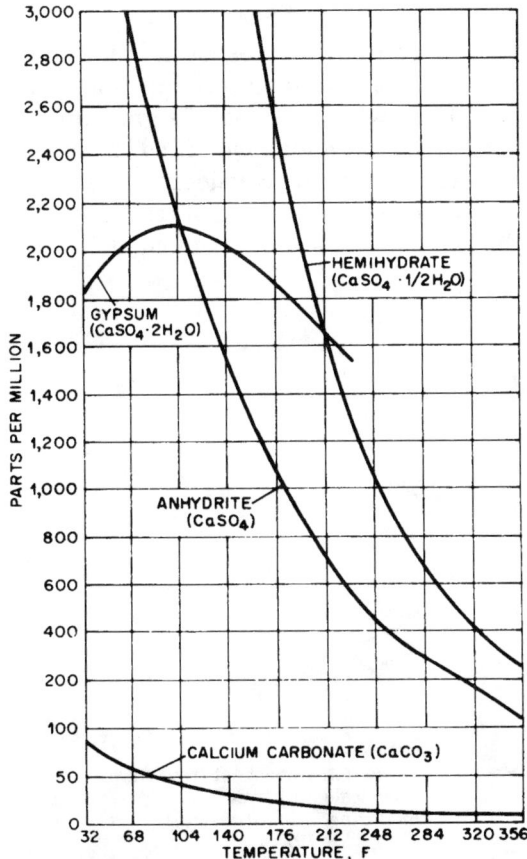

Figure 5.6 Solubility of calcium carbonate compared with calcium sulfate. (*Betz Laboratories, Trevose, Pa.*)

cult as controlling calcium carbonate scale. However, the common practice of applying sulfuric acid to reduce alkalinity so that higher cycles of concentration can be maintained in the cooling tower raises the sulfate content and may limit the cycles that can be obtained.

There are a number of formulas for calculating the maximum number of concentrations of calcium sulfate in cooling water before scaling occurs. A common one is as follows:

Maximum COC = 2400 mg/L hardness (as $CaCO_3$)/M (makeup) hardness

Much higher concentrations can be maintained by applying the phosphonate and acrylic polymer treatments. The increased ionic strength at higher concentrations also is responsible for increasing the solubility of calcium sulfate.

Inspection and maintenance of cooling tower surfaces

Yearly examination of heat exchanger surfaces exposed to the highest temperatures and subject to the most scaling is recommended for determining the effectiveness of chemical treatment. If significant scale is observed, cleaning with inhibited acid may be justified. If this is required more often than once every two years, consideration should be given to making a significant change in the method of treatment being applied. Figure 5.7 and Table 5.1 show the impact of condenser-tube

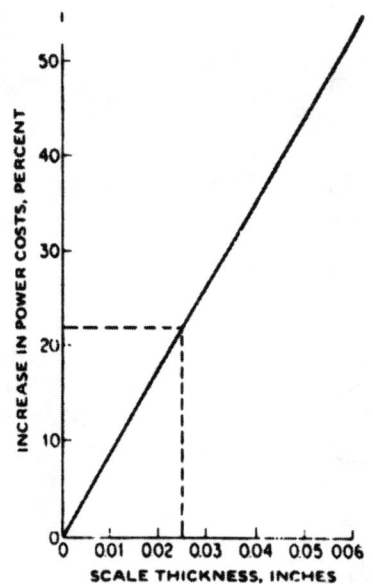

Figure 5.7 Impact of condenser-tube scale on power costs. (*Power magazine*, McGraw-Hill, New York, May 1972.)

TABLE 5.1 Heat Transfer Surface Required to Offset Fouling

Fouling thermal resistance, $(h)(ft^2)(°F)/Btu$	Overall heat transfer coefficient, $Btu/(h)(ft^2)(°F)$*	Thickness of scale† (approximate), in	Increase of required heat transfer area‡ (approximate), %
Clean tubes	850	0.000	0
0.0005	595	0.006	45
0.001	460	0.012	85
0.002	315	0.024	170
0.003	240	0.036	250

*The overall heat transfer coefficient U selected for this illustration is typical for a water-cooled refrigerant condenser. However, because it is possible to have different overall heat transfer coefficients depending on the systems, the effect of scale on the overall heat transfer will vary.
†Assume a mean value for the thermal conductivity of the scale of 1.0 $Btu/(h)(ft^2)(°F)$.
‡Square feet of inside surface of tube in heat exchanger.
SOURCE: *Heating/Piping/Air Conditioning,* April 1986.

scale on power costs and the heat transfer coefficient. The installation of apparatus for continuous monitoring of scale and corrosion is sometimes justified. This apparatus affords day-to-day observation as to whether scale-inhibiting treatment and bleedoff are adequate.

The author has observed cooling towers that had apparently never been maintained by proper water treatment and blowdown, with the result that scale almost filled up the space between the cooling tower packs of fill and was actually blocking airflow. A brief examination of this scale was done by subjecting it to a drop or two of hydrochloric acid. Effervescence due to carbon dioxide evolution was proof that the scale was likely primarily calcium carbonate.

Another simple test to apply to corrosion products is to expose black rust to a magnet. If the black rust is magnetic, it is probably black iron rust, or magnetite (Fe_3O_4). Red-brown iron rust is hematite (Fe_2O_3) and is more flocculent and nonmagnetic. What is the significance of this difference in observed iron rust? It is mainly that magnetite may provide a more continuous and effective protective iron oxide film to inhibit corrosion than the more flocculent red iron oxide, hematite.

Even with proper water treatment, it should be expected that the heat exchanger tubing will require brushing to remove thin dust-type deposits. This brushing should be done annually or less frequently.

Prevention and control of corrosion

Past practice allowed the use of a very effective zinc-chromate combination for corrosion inhibition; however, chromate has been banned by the U.S. Environmental Protection Agency (EPA) as a pollutant and health hazard. Corrosion inhibition is essential in cooling tower water, for it is aerated, contains high salt concentrations and suspended matter from hardness sludge formation and air contamination, and is maintained at warm temperatures—all factors that contribute to a corrosive environment. The corrosion of copper-bearing heat exchangers must also be given attention, usually by application of copper metal inhibitors, such as Item 147A sodium tolyltriazole.

MIC (microbe-induced corrosion) is getting considerable attention and may become serious if appreciable biological growths and crud are in the system. Sulfate-reducing bacteria (SRB), discussed more completely under the heading "Prevention of Algae and Slime Deposits," may become prevalent under crud or biofilms. It has been reported that a severe attack of MIC may result in as much as 1/16-in penetration of mild steel in a 6-week period. The degree of fouling (discussed later) has a direct bearing on the amount of MIC occurring.

Zeolite-softened makeup water provides a high pH in the concentrated cooling tower water and therefore requires a somewhat different method of corrosion inhibition. The high pH is beneficial in reducing corrosive tendency and reduces the amount of phosphate required for effective inhibition. At 100 mg/L total alkalinity, an orthophosphate (PO_4) test as low as 5 mg/L may prove adequate, while at 300 mg/L total alkalinity, PO_4 should be kept near 2 mg/L. Completely softened water (H = 0.0 mg/L) is not necessarily prescribed; however, appreciable hardness could cause calcium phosphate deposition if appreciable phosphate (PO_4) above 5 mg/L is present.

Commonly known *anodic inhibitors,* which control corrosion by inhibition at the anode, are chromate, nitrite, phosphonates, orthophosphate, azoles, and molybdate. *Cathodic inhibitors,* which control corrosion at the cathode, are polyphosphate and zinc. Modern cooling water treatment technology dictates that complex blends of both anodic and cathodic inhibitors, including dispersants, are necessary to achieve acceptable corrosion and scale inhibition. Some of the chemicals used for corrosion inhibition are discussed below.

Chromate. For many years, chromates provided very satisfactory corrosion inhibition; however, their use for comfort-cooling towers and their discharge to the environment have been banned by the U.S. EPA (see Table 3.1). Initially, concentrations of 500 to 1000 mg/L sodium chromate (as CrO_4) were found necessary for satisfactory corrosion inhibition. In later years, 10 to 100 mg/L sodium chromate (as CrO_4) with the addition of zinc salts and other supplements was found to yield proper corrosion inhibition.

Phosphonates. During the past 15 years, these chemicals have been found most acceptable as scale inhibitors; they have also been found to have reasonably good corrosion-inhibiting properties at higher concentrations (15 to 30 mg/L) in the concentrated cooling water than those required for scale inhibition. Improved corrosion inhibition is attained when phosphates are combined with azoles and acrylates and/or zinc. These higher levels of phosphonates may lead to the formation of calcium phosphonate and calcium phosphate scale. Limiting the calcium, the phosphonate, and the pH and increasing the acrylate levels of treatment are ways to control the calcium phosphonate, calcium phosphate, and zinc hydroxide and phosphate scaling. The phosphonates gradually degrade to form orthophosphate and create a scaling condition unless adequate acrylic polymers or the equivalent are included in the formulation.

Molybdate. The advantage of molybdate is that it is environmentally acceptable. The disadvantages are its high cost and less effective corrosion inhibition when compared with chromate. Supplementing molybdate with zinc, phosphonates, nitrites, acrylates, and azoles results in a very satisfactory treatment. Operation at higher pH levels of 8.0 to 9.0, rather than at normal lower pH levels of 6.5 to 7.5 with chromate treatment, has been helpful in making this less effective corrosion inhibitor adequately protective.

Phosphates. Orthophosphates, such as trisodium phosphate, acting as anodic inhibitors, are effective in the presence of oxygenated water (for example, cooling tower water); they fill in the voids in the formation of an effective protective film and accelerate its growth. Polyphosphates are less effective but better at neutral pH.

Zinc salts. Acting as cathodic inhibitors, zinc salts are effective supplements to the above chemicals; they generally provide corrosion reduction at the cathode by precipitation of zinc hydroxide. In addition, they counteract the formation of the copper-phosphonate complex, which can cause serious corrosion of copper-bearing metals and increased steel corrosion (metallic copper is deposited on the steel, and this results in galvanic corrosion).

Azoles. These chemicals, such as benzotriazole, mercaptobenzothiazol, and sodium tolyltriazole, serve to inhibit corrosion of copper-bearing metals by developing an effective protective film on the surfaces of these metals. They also provide a degree of corrosion inhibition for steel.

Acrylates. There are many varieties, such as polyacrylates, copolymers, and terpolymers, and they serve mainly as inhibitors to prevent the formation of scales composed of calcium carbonate, calcium phosphate, calcium phosphonate, and zinc hydroxide and phosphate. They enable the principal corrosion inhibitors to provide an effective protective film in the presence of potential scale-forming tendencies.

Coupons for monitoring corrosion control. In order to be assured that the corrosion control chemicals are serving adequately, steel and copper-bearing coupons should be installed for 30- to 90-day periods to determine the corrosion rate in mils penetration per year (mpy). The corrosion tester assembly and coupon holder and coupon (Figs. 5.8 and 5.9) should be located near the cooling tower so that discharge from the tester can be made to the cooling tower basin. The test method used is the ASTM D2688 Method A. Besides observation of the weight loss, the degree of pitting can be examined under the microscope when specimens are removed, since the pitting tendencies of the different treatments are of particular concern. A suitable time frame includes installing coupons in the spring, removing and replacing them in midsummer, and removing them in the fall when air conditioning is shut down. Assuming that water treatment is proper, it is to be expected that steel coupons will show not more than a 1.0-mpy corrosion rate (definitely below 5.0 mpy) and copper less than 0.3 mpy. Admiralty metal coupons should be installed in place of copper if heat exchangers are constructed of this material.

Table 5.2 shows the composition of components of various kinds of air-conditioning and refrigeration equipment.

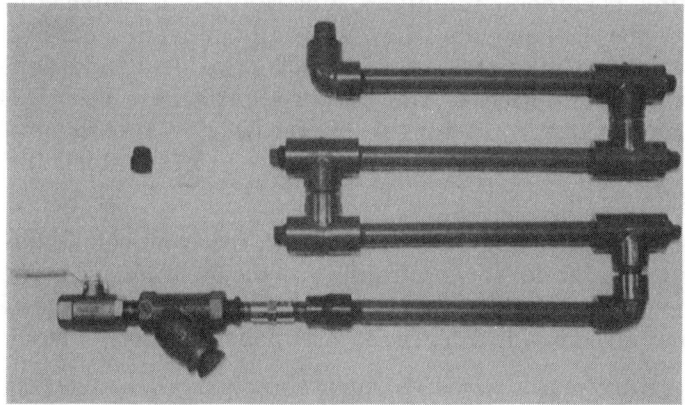

Figure 5.8 (*a*) Corrosion coupon rack. [*Frank N. Kemmer (ed.), The NALCO Water Handbook, 2d ed., McGraw-Hill, New York, 1988.*]

Open Recirculating Cooling Water Systems 133

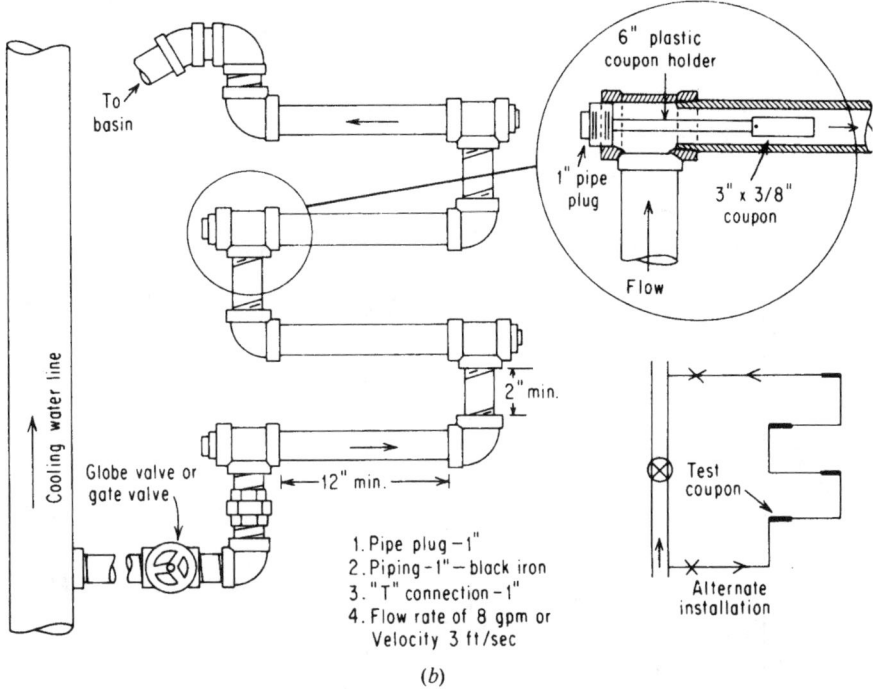

Figure 5.8 (b) Corrosion coupons. [*Frank N. Kemmer (ed.), The NALCO Water Handbook, 2d ed., McGraw-Hill, New York, 1988.*]

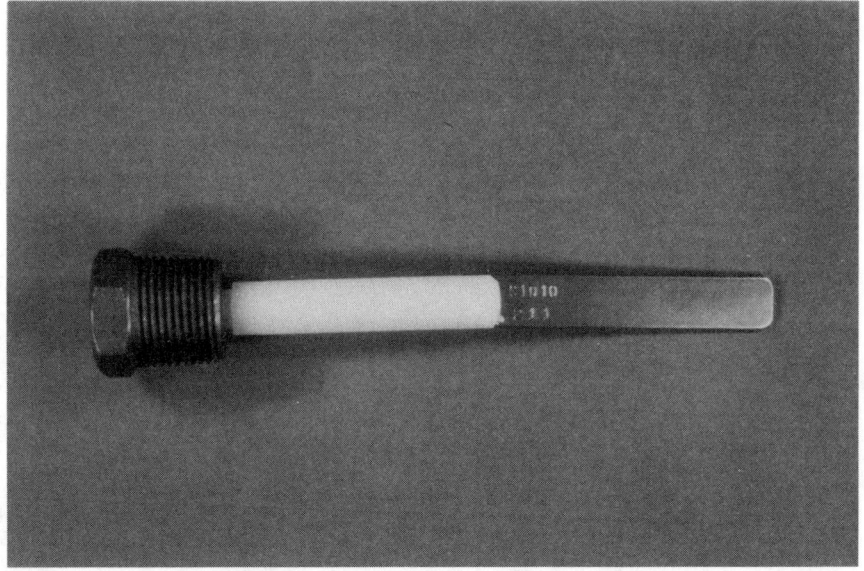

Figure 5.9 Corrosion coupon. (*R. W. Lane files.*)

TABLE 5.2 Composite Reference of Materials for Equipment Components

Equipment	Casings and support	Grids, fill, and eliminators	Pipes	Pans or basins	Ducts	Heat transfer surfaces	Valves and pumps	Fans
Cooling towers	Al S SS P	Al SS P W	Al S Cu P	Al S Cu Ct	SS W C			Al S Cu	Al S Ct
Evaporative condensers	Al S W Ct SS	Al SS Cu	P Ct	Al S Cu SS	Al S Cu	Al Ct SS	Al Cu SS	Al S Cu	Al S Cu	SS Ct
Cooling coils	Al S Cu SS	P W Ct		Al S Cu SS	Al S Ct		Al Cu SS	Al S Cu SS	
(Condensation)									
Air washers	Al S Cu SS	Al SS P Ct	Al S Cu SS	P Ct	Al S Ct	S	Al Cu SS Ct	Al S Cu	Al SS Ct
Room air conditioners	Al S Cu SS	Al Cu SS P	Al Cu SS P	Al S Ct	Al S Ct		Al Cu SS Ct	Al S Cu SS	P	Al Cu SS P	Ct
Air-cooled condensers	Al S W Ct Cu SS	Al S Cu SS	P W Ct	Al S Cu SS	P Ct		Al S Cu SS	W Ct	Al S Cu SS	Ct	Al S Cu SS	Ct
Heating coils	Al S Cu P(<65°C)	SS Ct Cu P(<65°C)	SS Ct	Al S Cu SS	Ct		Al S Cu SS	Ct	Al S Cu	

Al: Aluminum—aluminum in contact with water of high pH or containing certain dissolved salts, particularly those of copper, is subject to rapid localized attack. Special precautions in equipment design and water treatment may permit its use when this is desirable because of its low density or other considerations.
Ct: Coatings—galvanizing, plating, or protective coatings may be used over appropriate base material.
C: concrete; Cu: copper and alloys; P: plastics; S: steel; SS: stainless steel; W: wood.
SOURCE: *1987 HVAC Handbook*, American Society of Heating, Refrigeration and Air-Conditioning Engineers, Atlanta.

Details of methods of internal treatment.[11] An accepted and simple general treatment method* being used includes the proportional application of:

1. A calcium carbonate scale–inhibiting chemical like Item 95 diphosphonic acid (HEDP)
2. An organic dispersing and scale-inhibiting chemical like Item L218A, an acrylic terpolymer
3. Corrosion-inhibiting chemicals like Item L243 sodium molybdate or Item 7 sodium orthophosphate, and Item 18 tetrasodium pyrophosphate for inhibiting corrosion of steel and copper

At present, there are *four main water treatment programs* that have proved successful and require simple treatment control: molybdate, phosphate, zinc, and all-organic. The following dosage recommendations for these programs are based on operation of the cooling water at 4 cycles of concentration.

Molybdate program. The following composition is designed for effective scale and corrosion inhibition of cooling water:

7 mg/L (0.6 lb/10,000 gal) Item L243 sodium molybdate

20 mg/L (1.7 lb/10,000 gal) Item L234A polymer-phosphonate-zinc

1.5 mg/L (0.13 lb/10,000 gal) Item 147A sodium tolyltriazole

This composition (100 percent active) will be expected to provide about 10 mg/L molybdate (MoO_4), 10 mg/L acrylate, 16 mg/L phosphonate, 2 mg/L zinc, and 3 mg/L triazole in the concentrated circulating water (4 cycles). Tests required will be molybdate and conductivity to keep molybdate (Mo) at 5 to 8 mg/L and conductivity at 4 cycles.

Check tests at least once or twice a week for zinc, azole, and phosphonate are recommended.

Phosphate program. The following composition is designed for effective scale and corrosion inhibition of makeup water:

7 mg/L (0.6 lb/10,000 gal) Item L220A (acrylate and HEDP)

7 mg/L (0.6 lb/10,000 gal) Item L215C (trisodium phosphate, tetrapotassium pyrophosphate, and sodium tolyltriazole powder)

---

*In cases in which environmental regulations ban the discharge of orthophosphate or zinc, organics including Items L218A and 147A fed at higher dosages are reported to yield satisfactory scale and corrosion inhibition.

This composition provides 15 mg/L total phosphate (ortho and pyro), 3 mg/L triazole, 4 mg/L phosphonate (as PO_4) and 6 mg/L acrylate at 4 cycles of concentration of the makeup water.

Tests to be conducted and limits to be maintained are as follows:

Orthophosphate (PO_4): 3–6 mg/L

Phosphonate (PO_4): 2–5 mg/L

Conductivity: 4 × makeup water conductivity

The phosphate program is an effective corrosion-inhibiting program, but it requires careful control in inhibiting scale, particularly when used in the treatment of makeup waters of high hardness (>250 mg/L as $CaCO_3$) and alkalinity (>150 mg/L as $CaCO_3$). Acid treatment to reduce pH and alkalinity may be required to provide scale-free conditions and enable operation at the more economical cycles of concentration.

Zinc program. The following composition is designed for application to the makeup water to provide effective scale and corrosion control:

4–6 mg/L (0.4 lb/10,000 gal) Item L218A terpolymer (such as Rohm & Haas Acumer 3100)

3–4 mg/L (0.3 lb/10,000 gal) Zn as Item 35 $ZnSO_4$ or Item L247 $ZnCl_2$

6–8 mg/L (0.6 lb/10,000 gal) Item 95 diphosphonic acid

1 mg/L (0.08 lb/10,000 gal) Item 147A tolyltriazole

Sufficient Item L218A, 35, or L247 is recommended to provide 8 to 10 mg/L L218A, <5 mg/L zinc (Zn), 2 mg/L Item 147A triazole, and 16 to 18 mg/L Item 95 phosphonate (PO_4) in the concentrated cooling tower water as regulated by bleedoff at 4 concentrations of makeup water. Tests required for control are zinc, phosphonate, and conductivity.

In order to maintain zinc solubility, acid treatment is also required to lower pH and alkalinity and control calcium carbonate and zinc salts in solution. The U.S. Environmental Protection Agency (EPA) drinking water standard on zinc specifies a limit of 5 mg/L, and accordingly, discharge limits are specified at this same level. Applying 5 mg/L of zinc is not likely to result in excessive discharge in the bleedoff, since at normal cooling tower pH levels, some loss of soluble zinc occurs in the cooling tower owing to precipitation and adsorption.

All-organic program. The following composition is designed for application to the makeup water to provide effective control of scale and corrosion:

5 mg/L (0.4 lb/10,000 gal) Item 95 diphosphonic acid

4–5 mg/L (0.35 lb/10,000 gal) Item L218A terpolymer (such as Rohm & Haas Acumer 3100)

1 mg/L (0.08 lb/10,000 gal) Item 147 tolyltriazole

By the above application of Items 95, 147, and L218A, 12 mg/L phosphonate (PO_4), 7 mg/L Item L218A acrylate, and 2 mg/L Item 147A triazole will be provided in the concentrated cooling tower water as regulated by bleedoff at 4 concentrations of makeup water. Tests required for control are phosphonate and conductivity.

Simpler treatment methods involve purchasing a partially mixed combination of chemicals so that fewer chemicals are applied and application is less complex:

1. *Molybdate-acrylate-phosphonate-azole blend:* 0.4 lb/1000 gal of makeup of a combination of 60% of Item L230, 36% of Item L243, and 4% of Item 147A

2. *Molybdate-acrylate-phosphonate-azole-zinc blend:* 0.4 lb/1000 gal of makeup of a combination of 96% of Item L233 and 4% of Item 147A

3. *Acrylate-phosphonate-zinc blend with added molybdate and azole in bulk:* 0.4 lb/1000 gal of makeup of a combination of 48% of Item L234A, 48% of Item L243, and 4% of Item 147A

The advantage of these formulations is that they consist of minimum amounts of blended chemicals supplemented with molybdate and copper inhibitor in concentrated form (generic); therefore, they are less costly. The water treatment chemical companies have available formulations which contain these or similar chemicals and which are designed for any particular water supply. The user who wishes to operate more economically may (*a*) hire a consultant to advise on the particular water treatment chemicals needed, (*b*) obtain a specification for submission to various suppliers, (*c*) purchase the needed chemicals by competitive bidding, and (*d*) personally monitor the results or hire the consultant to continue supervision of the chemical treatment. There are many chemical blending companies from which one can purchase specified blends of water treatment chemicals.

Prevention and control of fouling caused mainly by algae and slime deposits

Fouling in cooling towers consists of complex mixtures of scale, sediment, biological matter, and oil. This non-scale-forming matter may be derived from iron or silt or naturally occurring organic matter in the makeup water, from particulate matter scrubbed from the atmosphere, and from migrating corrosion products formed in the system. Antifouling dispersants may be applied in order to penetrate and loosen the organic component and emulsify the oily layers. Such dispersants may be nonionic or cationic polymers or alkyl amines (such as Item L244, a dimethylamide).

Water treatment for air washers and spray coil condensers. These devices, often numerous and small-sized in some building systems, provide a real problem in controlling corrosion and scale. During the winter, when the air is dry and requires humidification, makeup water addition is high, while in the summer, sumps will likely be running over because of condensed humidity. Because of their small size and general nonimportance these devices generally get minimum attention. It is hoped that design has provided noncorrosive metals or plastics; however, treatment control is most difficult with the high and varied makeup water involved.

In the winter, generally high makeup water usage causes the conductivity of the concentrated water to be high; accordingly, increased treatment and a high bleedoff rate must be maintained to minimize scaling and corrosion problems. In the summer, the opposite is generally true: conductivity is low, makeup is minimal, and little if any bleedoff is required. As maintaining inhibitors at proper levels is difficult in the winter, application of Item 58, a slowly dissolving polyphosphate, may be the only feasible treatment method. In the summer, it is expected that the water will be so diluted by humidification that attempts to treat the water to maintain specific treatment levels will be almost futile. Neither scale nor corrosion should be a problem if the proper noncorrosive materials have been installed. Fouling and algae can probably best be controlled in both seasons by application of shot doses of the biocides listed later in this chapter. Periodic mechanical and inhibited acid cleaning may be expected to be required.

Prevention of algae and slime deposits.[17,18] Attention to the control of algae or slime growths in the cooling system is necessary to prevent these accumulations from occurring in the circulating system, as this could lead to stoppages, inefficient operation, Legionnaire's disease, and underdeposit corrosion. Monitoring bacterial counts, sulfate-reducing bacteria, and slime growths and applying appropriate quantities of biocides at proper locations and intervals are important functions for keeping the cooling tower in good operating order.

The cooling tower offers an ideal incubator for microorganisms, as it is operated at temperatures of 80 to 120°F (27 to 49°C) in the presence of sunlight, oxygen (air), and nutrients such as phosphorus, nitrogen, sulfate, and carbon dioxide. It is not surprising that algae grow abundantly in the sunlight, while bacterial slimes (anaerobic—not requiring oxygen and sunlight) grow abundantly in the heat exchangers.

There are three major classes of microorganisms that may exist in recirculating cooling water systems, namely algae, fungi, and bacteria. Algae range from unicellular plants to multicellular species; they contain colored pigments, such as chlorophyll; and they are known to flourish on wet surfaces in the presence of sunlight. Fungi are similar to algae, but they do not contain chlorophyll, and they require moisture and air but not sunlight. Bacteria are microscopic plantlike organisms; they lack chlorophyll and exist in three forms: rod-shaped, spherical, and spiral-shaped.

The biofilms that build up on piping and equipment provide an ideal environment for the proliferation of microorganisms. The biofilm causes a reduction in the heat transfer rate in heat exchangers and provides an ideal environment where sulfate-reducing and aerobic sulfur bacteria can develop acids and cause serious corrosion (MIC, or microbe-induced corrosion) of steel, as well as underdeposit corrosion. Biocides are added to kill the organisms or inhibit their growth.

Various *oxidizing* and *nonoxidizing* biocides such as those listed in Table 5.3, are now being used for controlling algae and slime growths. A discussion of these biocides follows.

Oxidizing biocides. The common oxidizing biocides are chlorine, hypochlorite, chlorine dioxide, chloroisocyanurate, and bromine. These are usually applied semicontinuously in dosages of a few ppm's, or they are applied in slug dosages of 1/4 to 1 lb or more per 1000 gal of water system volume several times a week, preferably to the distribution deck of the cooling tower. Application at night, when the sun cannot cause degradation of the biocides, may prove most effective.

Chlorine and hypochlorite are effective in algae control of cooling waters because of action of the hypochlorous acid formed when chlorine dissolves in water.

Chloroisocyanurate is more frequently used in swimming pool treatment, as it has the advantage of hydrolyzing slowly to release hypochlorous and cyanuric acids.

Chlorine dioxide does not form hypochlorous acid in water, but it is more effective at a higher pH, and it has the advantage over chlorine that ammonia does not detract from its effectiveness.

Bromine recently has been found to be more effective at a higher pH than chlorine. A convenient form of addition is Item L229, containing 1-bromo-3-chloro-5,5-dimethylhydantoin, which releases hypochlorous acid, hydrobromic acid, and hydantoin on hydrolysis in water. Using a combination, in a 1:1 molar ratio, of Item 47 12.5% sodium hypochlorite and Item L232 40% sodium bromide (equivalent to a ratio of 0.03

TABLE 5.3 Biological Control

Typical chemicals	Formula or abbreviation	Typical application	Limitations
1,3-Dichloro-5,5-dimethylhydantoin		Slow-release chlorination type. Sanitizer for swimming pools and small cooling systems.	Less active than chlorine, very expensive for large systems.
2,2-Dibromo-3-nitrilopropionamide	DBNPA	General biocide for low-pH cooling systems. Can be deactivated by increasing effluent pH.	Not applicable in high-pH systems. Expensive chemical.
Acrolein	$CH_2=CH-CHO$	General biocide effective in organic-contaminated waters. Deactivated by sodium sulfite.	Flammable, volatile, lachrymator. Dangerous to handle.
Bis (tri-n-butyltin) oxide	$(H_9C_4)_3-Sn-O-Sn(C_4H_9)_3$	Films out on surfaces, provides long-term surface protection. Effective on algae and fungi.	Very toxic to aquatic life, has severe discharge limitations. Foams.
Bis (trichloromethyl) sulfone		Wide-pH-range biocide for slime and algae.	
Bromine chloride	BrCl	Oxidizing biocide effective at high pH. Unaffected by ammonia.	More expensive than chlorine.
Bromine salts	$CaBr_2$, NaBr	Oxidizing biocide effective at high pH. Unaffected by ammonia.	More expensive than chlorine. Requires chlorine for production.
Calcium hypochlorite	$Ca(OCl)_2$	Chlorine type, dry chemical. Small cooling systems.	Adds calcium to the water. Expensive for large systems.
Chlorinated phenols		Highly effective, surface-active wood preservative.	Extremely toxic to fish, and generally objectionable. No longer available in United States.
Chlorine	Cl_2	Most common oxidizing biocide, effective for all biological control.	Difficult and dangerous to handle. Chlorine consumed by organics and ammonia.
Chlorine dioxide	ClO_2	Oxidizing biocide effective in neutral to high-pH cooling waters. Retains activity in ammoniated waters.	Requires hypochlorite and chlorine to make the chemical. Explodes at concentrations above 15%.
Copper sulfate	$CuSO_4$	Controls algae in ponds.	Not recommended for recirculating systems. Copper causes pitting of steel surfaces.
Dodecylguanidine hydrochloride	$n\text{-}C_{12}H_{25}NH-C(NH)-NH_3Cl$	Control of bacteria and algae. Protects wood. Surfactant.	Can form insoluble salts with phosphates and sulfates.
Methylene bis (thiocyanate)	$NCS-CH_2-SCN$	Broad-spectrum biocide. Controls anaerobic bacteria.	
Ozone	O_3	Sanitizing biocide for small systems. Electrically generated.	Low water solubility. Breaks down at pH above 8. Impractical for large systems.
Quaternary ammonium salts		Broad-spectrum biocide. Effective in high-pH systems.	May precipitate anionic polymers. Tendency to foam.
Sodium dichloroisocyanurate		Slow-release chlorination type. Sanitizer for swimming pools and small cooling systems.	Less active than chlorine, very expensive for large systems.
Sodium dimethyldithiocarbamate	$Na[(CH_3)_2N-CSS]_2$	Broad-spectrum biocide. Effective for pH 7 and up.	
Sodium hypochlorite	NaOCl	Chlorine type, liquid. For small cooling systems.	Expensive and increases pH. May cause $CaCO_3$ precipitation.

SOURCE: Thomas C. Elliott (ed.), *Standard Handbook of Powerplant Engineering*, McGraw-Hill, New York, 1989, pp. 4.235 and 4.236.

gal Item 47 to 0.02 gal Item 232 per 1000 gal of treated water) is an economical method of applying approximately 2 mg/L (theoretical) free bromine to large systems. A National Pollutant Discharge Elimination System (NPDES) permit may limit the discharge of bromine to 2 h per day and a maximum of 0.5 mg/L free bromine.

A periodic 2-mg/L dosage of L244, a hydrophobic biocide, is of benefit in providing penetration of biological growths or biofilm harboring sulfate-reducing bacteria (SRB) and in providing a means for chlorine (bromine) to be more effective. The author has had successful experience with this technique at a site experiencing a serious SRB infestation, which had caused corrosion of chiller copper tubing and costly replacements. Keeping algae growths in the cooling towers under control is also of benefit in reducing organic levels in the system and in reducing this food for biofilm and slime growth in the condensers.

Treatment with ozone, the strongest available oxidizing agent, has been receiving considerable favorable comment in recent years, even to the extent of reported effective corrosion and scale inhibition. It is a particularly good biocide, although it does not have a long life for later action in a circulatory system. While there are many reported cases of its providing excellent biocide action, eliminating blowdown, and providing corrosion and scale inhibition, the engineering and prediction of its corrosion- and scale-inhibiting abilities in new applications are still open to question. A recent paper[19] promotes its biocidal properties, saying it can be used along with traditional scale- and corrosion-inhibiting chemicals in the same way that chlorine is used with them.

It is acknowledged that ozone degrades the phosphonates, causing loss of scale-inhibiting ability, just as chlorine does. The author's opinion is that its use as a biocide should be strongly considered, but reliance should not be placed on its present reported success as a corrosion and scale inhibitor. Also, there should not be serious concern about its degrading the phosphonates, since scale- and corrosion-inhibiting treatments are added continuously; therefore, adequate phosphonate is always going to be present.

Another strong point for the use of ozone is that the biocides presently being used eventually may be banned for discharge, or their limits may be lowered even further. Using reducing agents, like sulfite applied before discharge, is a solution to this problem, but an additional treatment control problem is introduced. Ozone has the advantage of being an effective biocide at very low levels; also, its short life makes it ideal for discharge.

Nonoxidizing biocides. There are many nonoxidizing biocides, such as isothiazolones, quaternary ammonium and phosphonium salts, organosulfur compounds, methylene bisthiocyanates, sulphones and thiones, organodibrominated compounds, glutaraldehyde, and blends of these different biocides, which may be fed intermittently with chlorine once or twice a week to attain economical microbiological control. Since each cooling tower environment may be a little different, trial at varied dosages of many of the different biocides specified in App. A is suggested in order to find the most economical and effective dosage for the specific environment.

The isothiazolones, organodibrominated compounds, glutaraldehyde, and quaternary ammonium and phosphonium salts are not pH-sensitive like the methylene bisthiocyanates, sulphones, and thiones, which are effective only at a pH of 8 or below. It is reported that glutaraldehyde is an effective biocide at 10 to 100 mg/L dosage at a pH of 8.0. Since there is also concern that biofilms may develop in the heat exchanger and piping system, the periodic testing of the total bacterial count by dip cell testers* (Item L5242) is informative. Biofilms may be responsible for poor heat transfer; they may harbor bacterial growth and promote microbe-induced corrosion (MIC). Also, such monitoring will provide more assurance that *Legionella* bacteria are being kept under control so that aerosols from cooling tower drift will not pass this bacteria on to passersby.

Tests and precautions. Sulfate-reducing bacteria (SRB) are particular organisms which can cause serious corrosion and pitting but which can be satisfactorily controlled with bromine-hypochlorite treatment and effectively monitored with biological testers.† Since SRB may lodge under biofilms and crud and not be detected in the bulk water, sampling is recommended from a stagnant area of the cooling tower or from a biological monitor to obtain a better indication of whether there are any in the system. Cooling towers operate at an optimum temperature for bacterial growth, and since treatment with phosphonates and phosphates started some 15 years ago, algae growth problems have worsened because the phosphorus provides needed nutrient. As a result, algae growths provide food for other biological growths, such as anaerobic bacteria, and are apparently partially responsible for the general problem of increased biological growth.

A common way of feeding biocide treatment is to apply a shot-type dosage and then let the chemical level drop until the chemical reaches

---

*Easicult is one brand of an easy plant test for determining the bacterial count of cooling tower waters. It is available from Metalworking Chemicals & Equipment, P.O. Box 161, Chester, NJ 07930.

†Experience has indicated that the Hach Co. (P.O. Box 389, Loveland, CO 80539) Biological Activity Reaction Test (BART), No 24324 (Item L5241), serves very satisfactorily for the detection of sulfate-reducing bacteria.

its lower level of effectiveness. This is usually at 25 percent of the original concentration. The length of time required to reach the lower level may be estimated by the following formula:

$$T = 1.4(V/B)$$

where T = time between shots (retention time), days

V = volume of system, gal

B = blowdown and drift losses, gal/day

Calculations often indicate that application is required 2 to 4 times a week.

Monitoring and periodic removal of accumulations of crud in cooling tower basins is important. Such cleaning every 2 to 4 weeks is being performed by conscientious operators who wish to keep their systems clean and free of corrosion, scale, and the possibility of harboring *Legionella* bacteria. Another precaution to take to avoid problems from sulfate-reducing bacteria and other bacteria is to operate cooling tower circulation pumps for spare chillers at least 5 minutes a day when these chillers are off-line and not being operated. During this time, the currently effective biocide should also be applied. Circulating systems allowed to remain in a stagnant condition are often subject to bacterial growth problems.

Cathodic Protection

The close exposure of copper-bearing tubes to an iron tube sheet in a heat exchanger creates an ideal location for galvanic corrosion to occur. The installation of plastic-backed magnesium or zinc anodes on the face of the tube sheet has proved to be an effective way of combating this corrosion problem. Another way to attach the anodes to the tube sheet is by tapping and drilling the tube sheet so a 1/2-in-diameter stainless-steel 304 threaded bolt and a 1/16-in rubber gasket can hold the anodes to the tube sheet.

Legionnaire's Disease[17]

In July 1976, an outbreak of pneumonia at an American Legion convention in Philadelphia caused serious illnesses and deaths. The cause of this outbreak was later identified as a bacterium named *Legionella pneumophila*, which is commonly found in surface water supplies. Cooling towers are not the only source of *L. pneumophila*; however, the experience in Philadelphia gained national attention, as contaminated cooling tower water spray apparently entered the air intakes at the hotel, and hotel guests inhaled this mist or drift, which caused serious infections.

Since then, cooling towers have been considered a possible serious source of this pathogenic organism, although it is a common organism found in municipal water supplies, lakes, rivers, hot-water heating tanks, hot tubs, humidifiers, faucets, and showerheads.

In the May 11, 1992, issue of *Air Conditioning, Heating, and Refrigerating News,* it was reported that there were 6 dead and 100 infected in Sydney, Australia. So this disease is still with us and is probably the result of stagnant water, inadequate cleaning of cooling tower basins, and inadequate water treatment. In this case, suspicion has been cast on several cooling towers in shopping complexes, which have been considered the source of infection in the spring each year.

While there are many laboratories that can test for *L. pneumophila,* the National Centers for Disease Control in Atlanta maintains that money spent for such testing would be better spent in providing good housekeeping practices. *Legionella* outbreaks have not occurred in biologically controlled and well-maintained cooling towers. In cooling towers having high levels of bacteria, accumulated dirt, and biological debris in the sump, it has been observed that a significant *Legionella* outbreak is more likely to occur. The presence of other organisms seems to encourage the growth of *L. pneumophila.*

In order to keep this pathogenic organism under control in cooling towers, the following items should be given attention:

1. Mud and debris should be removed regularly from the cooling tower basin.
2. A cover should be installed above the distribution deck to prevent sunlight from fostering algae growth.
3. Dip cell tests for determining the total bacterial count should be conducted at least weekly, 3 to 6 h after biocide addition. Total bacterial counts should be kept below 10,000/mL.
4. Effective oxidizing biocides (chlorine, bromine types preferred) and nonoxidizing biocides should be applied intermittently. Treatment systems actuated by an automatic timer are particularly recommended.
5. Accurate records of the name of the biocide, the amount applied, the time applied, the dipstick reading, and the results obtained should be maintained.
6. In off-seasons, when the tower water lies stagnant in the basin, the circulation pump should be operated every few days, even when the chiller isn't being operated. Also, biocidal treatment and monitoring of this treatment should be given continued attention.

The personnel who clean debris from the cooling tower should wear protective clothing to keep from breathing aerosols containing the pathogenic bacteria.

The recordkeeping and the observations made after application of the biocide will be helpful in deciding on the type of biocide, the quantity of biocide, and the most effective time to apply the biocide when application is next required. It may even prove best to apply the biocide mainly at night, when sunlight isn't present to counteract the germicidal or oxidizing action.

The most effective biocides should be applied at the most effective dosage at proper intervals, as these chemicals are expensive, and overfeed is not acceptable for discharge to the environment. Although some literature does indicate that *L. pneumophila* may be resistant to chlorine, periodic (for example, hourly) application of chlorine, in addition to keeping the system clean and keeping bacterial counts under control, has been relatively successful.

Application of a number of nonoxidizing biocides, such as tributyltin, dibromonitriloproprionamide, glutaraldehyde, and methylene-bis-thiocyanate, has also been reported to be successful in controlling the growth of *L. pneumophila*. Some recently reported research reveals that cooling waters with total alkalinity above 250 mg/L or a pH above 8.5 are less favorable for *L. pneumophila* growth. While this doesn't mean that chlorine application or tower cleaning is unnecessary, the operator is at least more assured that the bacteriological control problem may not be as difficult in these types of cooling waters.[17]

Cleaning and disinfecting are recommended when the cooling system is first brought into service and also at the end of the cooling season or before a prolonged shutdown.

In order to be assured that *Legionella* bacteria are not prevalent in their guest rooms, hotels have learned to periodically add hypochlorite tablets to the stagnant water (condensate) dripping from chilled-water cooling coils in the room air conditioners.

Initial Conditioning of Cooling Water Equipment[20]

This information primarily applies to mild-steel equipment but may be applied to other materials as well. A pretreatment inspection is first prescribed to determine whether there is a need for treatment. Dirt on surfaces—dirt composed of oils, greases, or corrosion products—must be removed so that subsequent water treatment can be effective. New equipment also may likely require mild cleaning and passivating. *Passivation* is the initiation of a formation of a tightly bonded oxide layer that protects the metal surface; it is commonly accomplished by treatment with an alkaline polyphosphate surfactant.

The procedure normally used includes the following:

1. A cleaning-passivating solution composed of 90% Item 23 sodium tripolyphosphate or Item 18 tetrasodium pyrophosphate plus 10% Item L241 nonionic low-foaming surfactant
2. A passivating solution composed of water-soluble sodium chromate or dichromate
3. A cleaning solution composed of degreasing chemicals such as Item L241 (listed in 1 above)
4. A cleaning solution composed of sulfuric or sulfamic acid (such as Item 54) for derusting

Many of the water treatment companies have ready-prepared propriety chemicals available for accomplishing these cleaning and passivating steps. The normal dosage level is 0.1 percent at a pH of 7.8, and the prescribed circulation period at ambient temperature is 4 days or until the passivating film has formed. When the cleaning and passivating have been completed, a critical inspection of the equipment should be made, and if the equipment is not adequately cleaned and passivated, treatments should be repeated. The normal safety precautions, including the use of protective clothing, should be applied in handling chemicals. For more complete details on cleaning and passivating, see Refs. 14 and 20.

Cooling System Lay-up[21]

When cooling tower systems are out of service, steel water lines and copper-bearing heat exchangers may suffer serious corrosion from stagnant water in the lines. As the cooling tower systems are shut down, increased corrosion inhibitor treatment should be applied and circulated even in the lines to be exposed to freezing conditions. Then when these lines are completely drained, there will be a film of the corrosion inhibitor left in the lines and in the areas that drain incompletely. This residual will prevent serious corrosion. Of course, the areas that will freeze should be drained and dried as completely as possible. One of the bad effects of improper lay-up of cooling tower systems is the appearance of large, thin chips of iron oxide scale in the orifices of the distribution water deck when the tower is started up the following spring.

There are two different methods (dry and wet) of properly storing out-of-service equipment. It is usually most efficient to apply the dry method when freezing may be a problem. Just before either the dry or the wet method is applied, the blowdown should be increased so that the cooling water approaches the concentration of the influent makeup. The basin, the sides, and the distribution deck should be hosed down until they are free of sludge. If scale and fouling are in evidence, acid cleaning may be required for scale removal; then circulation and rinsing with a weak (1%) solution of soda ash is necessary following the acid treatment. After this neutralization, application of scale and corrosion inhibitors is required. Sodium tripolyphosphate should be applied at a dosage of 1 lb/1000 gal to attain a test of 100 mg/L phosphate (PO_4) and sodium tolyltriazole applied at 0.2 lb/1000 gal to attain 15 mg/L azole in the circulating stream for a period of several hours. Piping to be exposed to freezing conditions can then be drained. Sections of the piping and equipment that will not be exposed to freezing conditions should retain this corrosion-inhibiting solution in contact with metal surfaces subject to corrosion until they are returned to regular service and water treatment.

Before the tower is placed back in service, it should be thoroughly cleaned of any debris that has accumulated during the out-of-service period. Otherwise, these accumulations may cause fouling, UDC (underdeposit corrosion), and scale formation in the cooling tower and on heat transfer surfaces.

References

1. R. Burger, *Cooling Tower Technology,* 3d printing, Burger & Associates, Dallas, 1979.
2. D. Murphy, "Cooling Towers Used for Free Cooling," *ASHRAE Journal,* June 1991, p. 16.
3. K. R. Olsen, "How to Remove Fouling Solids from Cooling Waters," *Chemical Engineering,* May 1992, p. 155.
4. M. Brooke, "White Rust," *Materials Protection,* March 1992, p. 80.
5. *Betz Handbook of Industrial Water Conditioning,* 8th ed., Betz Laboratories, Inc., Trevose, Pa., 1980.
6. *Drew Principles of Industrial Water Treatment,* 1st ed., Drew Chemical Corp., Boonton, N.J., 1977.
7. Nalco Chemical Co., *The NALCO Water Handbook,* 2d ed., McGraw-Hill, New York, 1988.
8. R. W. Lane, *Industrial Water Treatment Guidelines and Water Analytical Methods,* Champaign, Ill., 1983.
9. *Cooling Water Treatment Manual,* TPC Publication 1, 3d ed., National Association of Corrosion Engineers, Houston, 1990.
10. *Pamphlets on Water Treatment Submitted to U.S. Air Force on USAF Contract FO7635-79-CO161,* Illinois State Water Survey, Champaign, Ill., 1979.
11. B. P. Boffardi, *Fundamentals of Cooling Water Treatment,* Calgon Corp., Pittsburgh, 1989.
12. *ASHRAE Handbook,* "HVAC Systems and Applications," "Water Treatment," American Society of Heating, Refrigeration and Air-Conditioning Engineers, Atlanta, 1987, chap. 53.

13. National Association of Corrosion Engineers, "Standard Recommended Practice On-Line Monitoring of Cooling Waters," *Materials Protection,* April 1990.
14. J. N. Tanis, *Procedures of Industrial Water Treatment,* chap. 33: "How to Preclean New Cooling Water Systems," Ltan Co., Ridgefield, Conn., 1987.
15. J. R. Rossum and D. T. Merrill, "An Evaluation of Calcium Carbonate Saturation Indexes," *Journal of the American Water Works Association,* February 1983, p. 95.
16. P. Puckorius, "Lecture on Cooling Water Treatment," Electric Utility Workshop, Champaign, Ill., Mar. 16, 1989.
17. R. A. Larson (ed.), *Biohazards of Drinking Water Treatment,* Lewis Publishers, Inc., Chelsea, Mich., 1989.
18. M. W. Mittelman and G. G. Geesey, *Biological Fouling of Industrial Water Systems: A Problem Solving Approach,* Water Micro Associates, San Diego, 1987.
19. S. L. Hollingshad and J. L. Sasfai, "The Effect of Ozone on Traditional Cooling Water Treatment Chemicals," Paper 348, *Corrosion/92,* National Association of Corrosion Engineers, Nashville, 1992.
20. *Initial Conditioning of Cooling Water Equipment,* National Association of Corrosion Engineers, Houston, 1982.
21. T. E. Gale and J. Beecher, "Suggested Guidelines for Laying Up Idle Equipment in Cooling Systems," *Proceedings of the 45th Annual International Water Conference,* Engineer's Society of Western Pennsylvania, Pittsburgh, 1984, p. 156.

Chapter

6

Closed Hot and Chilled Water Systems and Treatment

The modern building has many closed piping systems for transferring heat and cooling from boiler steam, hot water, electrically produced heat, and chillers to areas requiring the heat and cooling. These recirculating water systems are considered closed if the water is not exposed to the atmosphere and if makeup water requirements are very low.

The many types of closed systems are classified as chilled water (CW), ranging in temperature from 45 to 55°F (7 to 13°C) for air conditioning; medium-temperature water (MTW), ranging in temperature from 150 to 350°F (66 to 177°C) for heating buildings; and high-temperature water (HTW), for transferring heat above 350°F (177°C) in large complexes. CW and MTW systems are maintained under air pressure, while HTW systems are pressurized with steam or nitrogen.[1] An expansion tank is required to allow for the change in volume occurring when there are water temperature changes in the system. This may be a closed tank or an open tank located at the highest point in the system.

Definition of a Closed System

Technically, a closed system is one that does not use more than 5 percent of its volume in makeup (newly added water) in a year's time.[2–5] This means that a large building's chilled water system containing 100,000 gal of water shall not lose more than 14 gal of water per day. In the past, it was thought that closed systems, particularly chilled water systems, did not require water treatment because there was virtually no makeup water added and no opportunity for significant changes in the water composition within the system. While it is true that air (oxygen) does not enter these systems like it enters open systems (cooling towers) to cause corrosion and that hard makeup water is introduced only infrequently, corrosion and scale problems *do develop,* but more slowly.

Water losses do occur at pump seals or packing, and losses through overflow of expansion tanks is likely caused by the original design providing undersized equipment that is inadequate for handling the expansion and contraction of varying loads. Surveys show that losses of 25 percent of system volume per month are not uncommon and that appreciable water losses are the rule, not the exception. It should be realized that what appears to be a small leakage or loss can have long-term significance; for example, a pump gland leakage of 1/2 pt/min actually amounts to 32,000 gal/year. This can account for the production of 2 1/2 lb of dry rust, which will occupy almost 0.1 ft^3 (nearly 3000 cm^3) somewhere in the piping system.

The efforts required in controlling water loss can be frustrating, and there are cases in large, complex facilities in which finding where the losses are occurring becomes almost impossible. The gross loss in such cases is probably derived from a multitude of small leaks or losses.

Older buildings used hot water or steam radiators for heating at temperatures of 150 to 200°F (66 to 93°C) at which there is a low level of oxygen solubility; large radiators provided venting of air. These larger lines and radiators also provided considerable space for corrosion product buildup, which was relieved at steam traps and strainers. Present-day closed hot water and cooling systems for transferring energy include thousands of feet of small-diameter piping, hundreds of small control valves and temperature-sensing devices, and dozens of pumps that may have leaking packing glands.

Since it is common to have dissimilar metals in these systems, galvanic corrosion is often a problem if makeup water additions are excessive. Keeping the corrosion inhibitor concentration at an effective level is essential and may require extra effort. The addition of excessive makeup water, which contains minerals and dissolved gases, such as air (oxygen), increases the corrosive and scaling tendencies appreciably in these systems.

Corrosion and Its Control

Corrosion is stifled in the absence of oxygen, and scale will not form if added makeup water containing potential scale-forming ingredients is kept at a prescribed minimal level. Calcium carbonate is the most common scale formed but may not likely become a problem unless a continuous loss of water is allowed to occur. The corrosion products of iron and copper are usually the major foulants. These foulants can cause:

1. Erosive damage to pump packing and mechanical seals.
2. Clogging in piping and strainers.
3. Decreased heat transfer efficiency, resulting in a reduction in the system's capacity to heat or cool.
4. Controls to become inoperative.
5. Corrosion in chiller tubes, in the form of pitting.
6. Underdeposit corrosion and biological growths when they lodge in minimum-velocity areas. (In one case, in which a closed system was not treated, iron rust from the steel piping system lodged in the chiller on the copper heat exchanger coils and caused pitting and perforation to occur underneath the rust film within a few years.)

To better appreciate the significance of small leaks or losses in a closed system, let us assume that there is leakage at the rate of 1 gal/h (about 30 drops per minute). Under these conditions, it was calculated that enough dissolved oxygen can be introduced into the system by the new makeup water (containing the normal 10 mg/L dissolved oxygen) to form sufficient rust to solidly clog 14 ft of a 1-in steel pipe.

Fouling and Its Control

Chemical treatment should be initiated when systems are new and clean, as it is difficult to clean and treat closed systems after they have become fouled.

Bacterial fouling may cause slime or a biofilm to form on tubes, thus reducing heat transfer. Such fouling may also cause the formation of black, foul-smelling corrosive water of low pH (acidic), and it may foster the growth of sulfate-reducing bacteria (SRB). Proper biocide treatment to prevent the growth of SRB is essential, since SRB are capable of initiating serious pitting-type corrosion and subsequent perforation of copper tubing and steel piping.

Accurate monitoring of closed systems for water losses is particularly necessary to properly control scale and corrosive tendencies. Metering the influent water and monitoring the water treatment concentration are important steps. Since water meters may be inaccurate under conditions of very low flow (seepage), solenoid or level control valves actuated by an appreciable water level change should be installed ahead of the water meter. Then more accurate flow information can be obtained from the water meter during these periodic higher flow conditions.

A recommended technique for properly monitoring makeup water usage and water treatment is to take a sample of the water in the system and test the corrosion inhibitor concentration in the sample. This should be done at least monthly (preferably weekly), and the treatment should be adjusted as needed. Additional advantages of this technique are as follows:

1. The percentage of water losses from the system may be estimated from the noted lowering of the test results and a check on water meter readings.
2. An estimate of the volume of the water in the system can be calculated from knowledge of the pounds and strength of the chemical added to the system, making possible more accurate chemical treatment control.

This method of determining the volume of the water in the system may likely be more accurate and less tedious than calculating the volume from the piping size and the length of the many different piping systems in the building. This technique is illustrated in the following example and in the use of Table 6.1.

Example If 5 gal (weighing 50 lb) of Item 62 hot water boiler blend containing 67% sodium nitrite (NO_2) solution was added to a system and the sodium nitrite

TABLE 6.1 Nitrite Tests vs. Water Loss (Percentage Calculation)

Percentage of initial concentration	Percentage loss of system volume
100	0
95	6
90	12
85	17
80	22
75	30
70	37
65	46
60	55
55	64
50	72
45	83
40	94
35	108
30	122
25	143
20	163
15	190
10	230
5	300

Closed Hot and Chilled Water Systems and Treatment 153

test of the system water was 500 mg/L (as NO_2), equivalent to 500/0.67 or 746 mg/L sodium nitrite, the volume of the system would be calculated as follows:

$$50 \text{ lb} \times 0.67 \times 0.67 = 22.5 \text{ lb sodium nitrite}$$

$$746 \text{ mg/L} = 746/1{,}000{,}000 = 22.5/x$$

$$746x = 22{,}500{,}000$$

$x = 30{,}161$, then divided by 8.34 (to convert pounds to gallons) =

$$3616 \text{ gal (volume of system)}$$

If the nitrite in this system had tested 600 mg/L nitrite (as NO_2) at the start of the month and 400 mg/L at the end of the month, calculations show that the final concentration would be 400/600 × 100, or 67 percent of the initial concentration. Table 6.1 indicates that this is a 40 percent loss of the initial volume—a serious loss.

Since a closed system has been defined as one that loses less than 5 percent of its volume annually (0.42 percent/month, 0.014 percent/day), this particular system is operating with too much loss, approximately 8 times too much. If this amount of loss had occurred in 1 year instead of 1 month, the loss of the initial volume would have been 3.3 percent, which is considered satisfactory for a closed system. If this amount of loss is allowed to continue, serious scale and corrosion problems will likely develop. Keeping a record like that shown in Table 6.2 proves helpful in controlling the makeup usage and water treatment.

Using the following equation is another method for estimating the volume of water in a closed system.

$$V = (W)(T)(1200)/(C_2 - C_1)$$

where W = the pounds of chemical added to the system
T = nitrite, % as NO_2
C_2 = mg/L nitrite (as NO_2) after addition of chemical
C_1 = mg/L nitrite (as NO_2) before addition of chemical

TABLE 6.2 Record of Makeup Usage and Treatment

Date	Meter reading	Water used, gal	Chemicals added, lbs	Tests, mg/L	% Loss
10/12/90	100000	0	50	500	0
10/19/90	101000	1000	0	475	5

In this 20,000-gal system, 1000 gal, or 5 percent, was lost in 1 week, equal to 260 percent in 1 year—much above the requirement for a properly controlled closed system.

System volume may also be estimated from the size of the expansion tank. On the basis of standard HVAC design tables, these tanks are designed to be half full when initially filled at ambient temperature and three-quarters full at operating temperatures in MTW and HTW systems.

Source of Leakage

Makeup should be applied through a U.S. EPA–approved makeup system with an air gap or a backflow prevention device. This is necessary, as the water in a closed system is nonpotable (unfit for human consumption), and the U.S. EPA Drinking Water Regulations prohibit cross connections between nonpotable and potable water systems. Installation of an air gap in the makeup water line or a reduced pressure zone backflow preventor is required.

As mentioned previously, periodic testing (at least monthly, according to the method described in App. C) of the nitrite content of the closed system water should be conducted to check on the water loss and makeup water usage. If continued high nitrite consumption is required to maintain the recommended nitrite test results, it may be concluded that serious losses are occurring. Follow-up action is recommended to locate the source of the leakage and make immediate repairs. The losses may likely be at these locations:

1. Expansion tank or makeup station, where overflow may be occurring or valve repair or redesign required
2. Multiple makeup water connections
3. Air vents or eliminators
4. Connections with other piping systems
5. Pump packing or mechanical seals

Chilled water derived from electrically driven compression-type or steam-absorption type-refrigeration equipment (chillers) may likely be supplied at temperatures of 45 to 55°F (7 to 13°C) to heat exchangers, where the circulating water may be tempered with warmer water to attain the required temperatures for circulation to the individual building areas. In colder climates, hot water may be generated at temperatures of 120 to 150°F (49 to 66°C) or at 300 to 350°F (149 to 177°C) in HTW systems and circulated to heat exchangers for blending to attain the required temperatures for building comfort. Now modern multistory buildings even in medium-cold climates may require only cooling during the whole year, since the building itself retains heat and the process equipment, lighting, and building occupants contribute appreciable heat.

All closed systems require expansion tanks to allow for volume changes due to temperature changes. Inadequate expansion tank capacity often is responsible for water losses. Expansion of the system water during daytime operation (high-usage periods) and contraction during the nighttime (nonusage periods) causes extreme differences in the required total water volume of the system; as a result, expansion tanks may overflow. It is now common to include nitrogen as the atmosphere in expansion tanks in order to avoid the introduction of air (oxygen) from the outside atmosphere into these systems. However, additions of new makeup water, which is saturated with oxygen (6 to 12 mg/L is normal; see Fig. 4.5), will likely supply sufficient oxygen to keep the corrosion process functioning. This makes avoiding excessive makeup water additions even more important.

Thermal Energy Systems

Ice and chilled water are produced in thermal energy systems (TES) during off-peak periods of electrical production to reduce electric current costs during peak periods. These systems, which are also desirable because they provide stored cooling capacity, are coming into increasing use to provide savings in electric energy costs. Corrosion (and biological activity) may still occur in these closed systems (some of which have a volume of 1 to 2 million gal) at low temperatures of 36°F (2°C); therefore, the addition of corrosion inhibitor and biocide treatments is still required. It has been proposed that fire protection systems and thermal energy systems be combined into one system, although it is not known whether this has yet been done practically.

Free Cooling[6,7]

In free cooling systems (better called ambient systems), cooling tower circulating water may serve as the water supply for the building closed system. This practice, as described more fully in Chap. 5, provides energy savings in the spring and fall, when the cooling load is lower. It has the advantage of eliminating chiller operation and therefore providing significant energy savings.

Since cooling tower water is saturated with oxygen and probably has a high mineral content owing to evaporation in the cooling tower, it is more corrosive; therefore, increased attention to the chemical treatment and monitoring of the closed system will be required. The chemical treatment and treatment levels maintained for cooling towers are not designed for closed systems, in which the corrosive environment may be more severe owing to stagnant, low-flow conditions. In some cases, to reduce this increased corrosion risk, heat exchangers (Figs. 6.1 and 6.2) are installed so that the cooling tower water can provide the desired cooling of the closed loop without being an actual part of the circulation in the closed system.

Glycol System[8]

The addition of glycol (ethylene or propylene) to the extent of 15 to 30 percent is often made to closed systems to provide freeze protection for preventing possible piping rupture. Glycols are oxidized in the presence of air and consequently form organic acids; as a result, they become corrosive if corrosion inhibitors are not added. The corrosion inhibitors buffer the organic acids so that they are not corrosive. Supplemental inhibitor may be added so that glycols may last 12 to 18

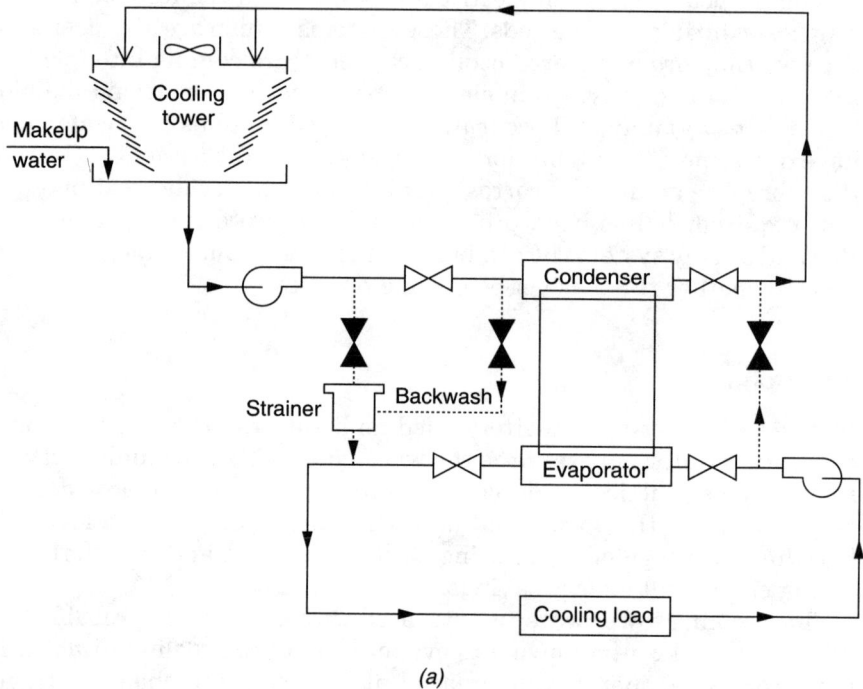

Figure 6.1 Free cooling. (*a*) Isolated mode. (*Courtesy of K. A. Selby, Puckorius & Associates, Evergreen, Colo.*)

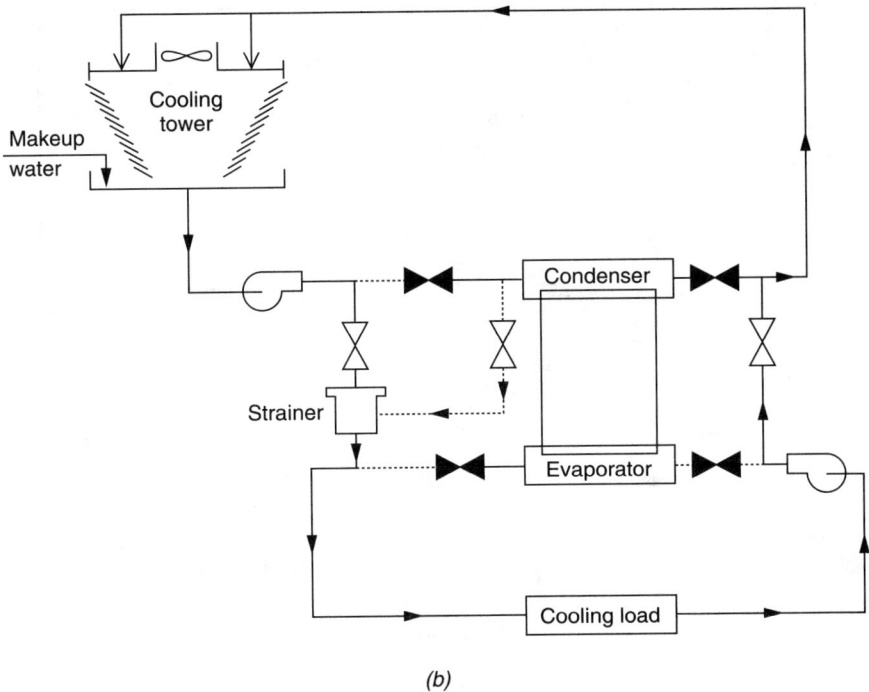

Figure 6.1 Free cooling. (b) Direct injection. (*Courtesy of K. A. Selby, Puckorius & Associates, Evergreen, Colo.*)

years if properly maintained.* Propylene glycol combined with its inhibitor dipotassium phosphate is low in acute oral toxicity and thus is accepted by the U.S. Department of Agriculture (USDA) for use in heat exchangers in food plants. In general, the glycols may be used for contact with steel, cast iron, copper, brass, solder, and plastics but not with galvanized steel above 100°F (38°C) or aluminum above 150°F (66°C).

Glycols should be completely drained from closed systems before summer chilled water operations begin, since they can cause increased bacterial growth by serving as nutrients. Biocide treatment is often required when summer chilled water service begins.

---

*Dow Chemical Co., Midland, Mich., a manufacturer of propylene and ethylene glycols, provides a free analysis of its products for concentration, pH, reserve alkalinity, freezing point, and appearance if the user has a glycol system with a capacity greater than 250 gal.

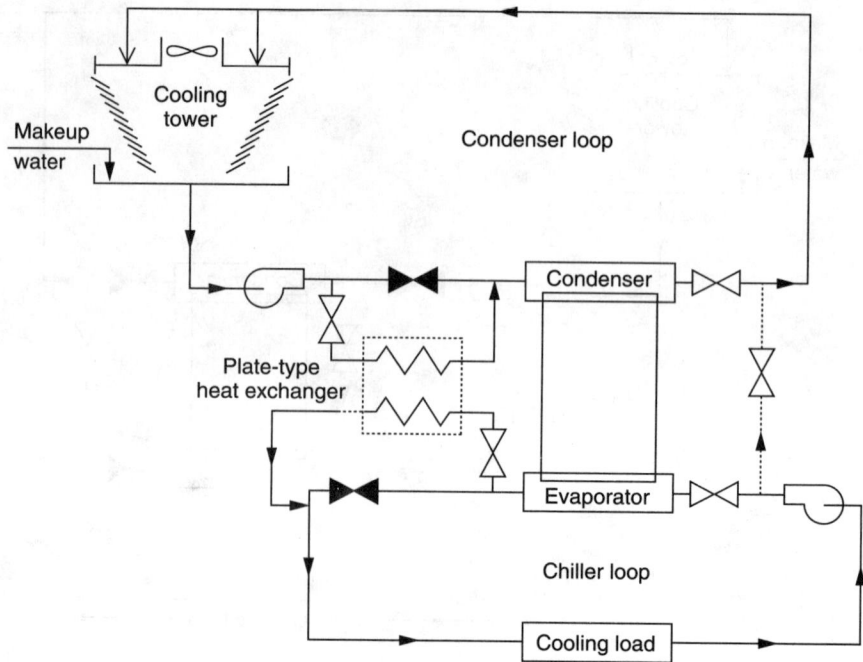

Figure 6.2 Free cooling, isolated with plate-type heat exchanger. (*Courtesy of K. A. Selby, Puckorius & Associates, Evergreen, Colo.*)

Hot Water Closed Systems

In addition to MTW and HTW systems, there are two kinds of hot water heating systems: fired and unfired. The fired system supplies steam or hot water from a small steam or hot water generator directly to the point of use. The unfired system supplies heat to the circulating water system by means of a heat exchanger, which is supplied with heat from a central heating plant. Hot water closed systems may be expected to experience more maintenance problems than the chilled systems because of operation at higher temperatures [120 to 150°F (49 to 66°C), or to 350°F (177°C) for HTW systems]. Therefore, the need to provide chemical treatment and maintain it at specified levels is even more acute.

HTW systems are pressurized with steam or nitrogen (preferably nitrogen) and provide heating by pumping the pressurized water to the outlying facilities. Advantages are that a deaerator, steam traps, and condensate tanks are not required and water treatment control is simpler. Normally, makeup water is zeolite-softened, and the treatment consists of applying caustic soda to provide a pH of 9.5 to 10.5 and sodium sulfite to provide 50 to 100 mg/L (as SO_3).

Chemical Treatment

There are a number of different treatment programs that are prescribed for closed systems. Treatment is usually not designed for scale control, because the makeup water usage is assumed to be so low that scale should not be a problem. If it is known that the makeup water to be added to the system has appreciable hardness, the recommendation is usually given to install a sodium ion exchanger (zeolite softener) so that softened water can be applied as makeup. Phosphonate or polyphosphate treatment is not considered adequate, because these chemicals degrade during long periods without treatment additions.

Assuming that the quantity of makeup is very low, the amount of scale formed from use of a moderately hard water is minimal and not usually sufficient to be of concern. The initial fill may provide a minimal degree of scale but then the potential for forming appreciable scale is exhausted by the softening (hardness precipitation) and therefore is of no further concern. Corrosion inhibition of the metals in the system, mainly steel and copper, is of major concern.

The methods of treatment are as follows:

pH control

Maintaining a minimum pH of 9.0 (the acceptable pH range is 9.0 to 10.5) usually may reduce corrosive tendencies to acceptable levels. This means that application of alkaline chemicals, such as caustic soda, soda ash, or trisodium phosphate, will be required to raise the pH. Except in cases of possible cross-connection contamination of the potable water system, this simple method of corrosion inhibition is not the generally recommended method of treatment because of the scale-forming potential and because better and more all-encompassing methods are available.

Chromate inhibitors[9-11]

While chromate treatments are exceptionally good corrosion inhibitors for closed systems, the U.S. Environmental Protection Agency (EPA) has specified that chromates not be used in comfort (open) cooling tower systems. While this ultimatum may not apply to closed systems, discharge at any time would be severely reprimanded, and even having chromate in stock may be equally condemned. Because of the risk involved in having a leak and discharge, the use of chromate is not recommended. If found acceptable by the U.S. EPA for use in closed systems in unusual cases, sodium chromate or dichromate should be applied at 250 to 350 mg/L chromate (as CrO_4) at a pH of 8.0 to 9.5 in the circulating water. It is a very effective corrosion inhibitor and has the advantage of being sufficiently toxic to counteract microbial growth in the system water. Its disadvantages are its incompatibility with antifreeze products and its corrosive effects on mechanical pump carbon seals* when used above 500 mg/L chromate (CrO_4).[11,12] Sufficient soda ash or caustic soda should be applied with the chromate to satisfy the pH requirement for the overall treatment.

Oxygen-scavenging inhibitors

These sulfite and hydrazine inhibitors[11] (preferably catalyzed) are desirable for their ability to reduce oxygen concentration and consequent corrosion. Sodium sulfite may be preferred to hydrazine because of the toxicity of hydrazine, but it does have the disadvantage of providing sulfate derived from oxidation of the sulfite; the sulfate promotes the growth of sulfate-reducing bacteria. Oxygen-scavenging chemicals are not considered generally satisfactory chemicals for application to closed systems, since continued addition and frequent testing are necessary to maintain the proper control to match the oxygen ingress into these systems.

Nitrite-azole inhibitors[13–16]

These blended products serve generally as the most accepted products for corrosion inhibition of steel and copper-bearing metals in closed systems. This is particularly true in systems employing antifreeze or when the use of chromate is not considered. Azole inhibitors, such as the sodium salts of mercaptobenzothiazol, benzothiazole, and tolyltriazole (Items 112, 113, and 147A), are applied along with nitrite to provide inhibition for copper-bearing metals, and borax is applied as a buffer chemical to provide a desired pH of 9.0 to 10.5 in the system water.

Usually, a blended dry product (Item 62; see App. A) containing sodium nitrite, plus borax and soda ash and Item 112 sodium mercaptobenzothiazol, is specified to provide the necessary inhibitor concentrations in the water system. The purpose of the nitrite, borax, and soda ash is to inhibit the corrosion of the steel, and the purpose of the mercaptobenzothiazol is to inhibit the corrosion of copper-bearing metals in the system.

---

*Replacement of carbon seals with ceramic seals is suggested.

Maintenance of 500 mg/L (as NO_2) or the addition of 10 lb of Item 62 per 1000 gal of system volume[12,13] is prescribed for control of corrosion inhibition in the system and is used as the test to provide the necessary control of treatment. Contrasted with chromate, which causes wear of mechanical pump carbon seals above a 500-mg/L dosage, nitrite treatment causes no wear up to a 4000-mg/L dosage (as NO_2). Claims are made that nitrite plus inorganic phosphate inhibitor blends are superior to nitrite or phosphate alone.[14]

A gradual decline in nitrite concentration and the formation of excessive gas in the system are indications that bacteria are probably thriving in the system. The addition of a biocide (N-alkyl trimethylene diamine acetate) or Item 108 (a biocidal reagent blend containing a quaternary amine-organo-tin mixture) may be necessary to counteract the growth of denitrifying (*Nitrobacter agilis*) bacteria.[14-16]

Difficulties still can be experienced in removing these bacteria from a system because the biocide must reach all segments of the system, and there may be dead ends or other areas where the biocide does not contact. The result is reinfection of the system when start-up is resumed. Therefore, precautions must be taken to ensure that all parts of the system are contacted by the biocide applied.

Silicate inhibitor[17]

Adding Item 32 liquid sodium silicate to increase the silica (SiO_2) to 20 to 50 mg/L and maintaining the pH in the 7.5-to-9.5 range provides effective corrosion inhibition of steel and copper-bearing metals. The disadvantages of silicate inhibitor may be the development of a thin insulating scale layer, the length of time required to develop effective corrosion inhibition, and the dispersive reaction with iron oxide to form a dirty-appearing water. Advantages are that it is nontoxic and does not contribute to bacterial growth.

Molybdate-azole inhibitors[18]

A formulation consisting of 150 mg/L sodium molybdate (Item L212 or L243) (as Mo), 10 mg/L Item 147A sodium tolyltriazole, and an alkaline chemical (Item 5 soda ash) to provide a water system pH of 8.5 to 9.0 has proved to be very satisfactory for corrosion inhibition. Item L246 contains these ingredients in a liquid mixed formulation, but it is more costly in this form. Other formulations that include small percentages of Item 95 phosphonate, Item 35 zinc sulfate, and Item L218A acrylic copolymer may have some advantage in providing slightly better scale control and corrosion inhibition of steel. There is no indication that biological growth problems are experienced with the molybdate formulations, but if such problems were encountered, application of a biocide (Item 108 or Item 131) would be expected to be effective. The disadvantage of the molybdate formulations are their high cost. Their advantages are that they are nontoxic, they do not support bacterial growth, and they are compatible with antifreeze (glycol) products.

Low-Pressure Steam or Hot Water Boilers

The nitrite-borax treatment mentioned above (Item 62) or a liquid version thereof is also an effective treatment for low-pressure steam or hot water boilers. The procedure of maintaining 500 ppm nitrite (as NO_2) in the Item 62 formulation in hot water systems has been used successfully for years to provide the necessary corrosion inhibition.

Corrosion Testing

If the building manager is to keep abreast of the condition of piping and equipment in the building, installation of corrosion coupons at a testing station should be arranged. This would be similar to the pipe insert corrosion testing described in Chap. 2 and the coupon corrosion testing described for the cooling tower water in Chap. 5 (see Figs. 2.3, 2.4, 2.5, 5.8, and 5.9). Exposure of inserts or coupons for 2- to 6-month periods will provide the necessary information for determining whether present piping and equipment are being properly cared for or whether changes in operation or treatment are necessary.

Cleaning of Systems

New systems should be *flushed thoroughly* to remove pipe mill scale, cutting oil, corrosion products, metal chips, pipe joint compound, solder flux, welding splatter, dirt, and other debris before being placed in operation. Reference 19 provides a complete description of the correct ways of flushing and chemically cleaning new and old closed systems. It is suggested that readers refer to this text, particularly if they intend to do the cleaning with their own forces or if they plan to write up specifications for the project. Such cleaning is a complicated process when done effectively and should undoubtedly be placed in the hands of a chemical cleaning or water treatment company that is experienced in this field. In many older buildings, the systems never received water treatment, and so they have become partially plugged with corrosion products. Corrosion products in general are difficult to remove, and if they are subjected to strong solvents, such as acids, the result may be piping leaks and property losses. Also, there are hard iron oxide flakes (magnetite), which are particularly abrasive to pump parts and impellers, but these flakes can be effectively removed by magnets that are installed in strainers. Besides the conventional chemical methods offered by the chemical cleaning companies, there are some newer techniques that have been described in recent papers[20,21] presented at water conferences. Reference 21 cites a neutral pH process in which phosphonates and an iron-sequestering polymer are used in the cleaning process.

Complete piping replacement, though costly, may prove to be the best solution. Of course, the cause of the corrosion should be determined so that replacement is made with a material more resistant to corrosion. Also, the new system should be properly cleaned, and installation of a proper water treatment system and a means of monitoring the new system should be arranged.

Sidestream (5 to 10 percent of full flow) and full-flow filter systems are often installed to remove suspended solids and are effective particularly during system start-up. Suspended matter in the circulating water may cause deposits to form, which may lead to the development of underdeposit corrosion cells and deep pits. In closed systems that require appreciable makeup water owing to losses, in-line filters capable of removing particles down to 1 µm in size have been found necessary to prevent plugging of the small-diameter lines and valves preceding room induction coils.

When the techniques described by Rosa[19] are used, the following formulations can be applied effectively at a dosage of 1.0 to 1.5 lb/100 gal to remove foulants and deposits. These formulations should be dissolved in 5 to 10% water solutions (20 to 40 lb/50 gal) in the tank used for water treatment chemicals. Then the solutions can be circulated for several hours in the system with continuous bleedoff.

(It is important to apply the usual safety procedures in handling these full-strength chemicals, and care should be taken not to introduce the chemicals into potable water connections or into associated equipment which may be vulnerable to chemical attack.[19])

Caustic soda–phosphate blend, consisting of:

Item 4 caustic soda, 15%

Item 7 trisodium phosphate, 5%

Water, 80%

Polyphosphate, consisting of:

Item 23 sodium tripolyphosphate, 10%

Water, 90%

Polymer blend, consisting of:

Item 95 diphosphonic acid, 10%

Item 218A acrylic terpolymer, 10%

Item 4 caustic soda, 6%

Softened water, 74%

Complete draining of water from the system at the lowest point should be provided during the flushing process. Then, when the flushing process is resumed, complete removal of the cleaning materials should be assured. This may be checked by testing the water with phenolphthalein indicator (App. B), which should react "colorless" rather than pink when the pH is below 8.2. If complete assurance is desired regarding whether the closed system is completely protected from corrosion, then corrosion coupons or nipples, as described in ASTM D2688 Method A or B (see Chaps. 2 and 5), should be installed. Corrosion test results should be maintained below a corrosion rate of 0.5 mpy.

Fire Protection Systems

Wet fire protection systems are filled with water and not ordinarily emptied and refilled more often than once a year according to practices in downtown Chicago. Corrosion may not be a real problem in plain steel pipe in these systems, as oxygen in the fill water is soon depleted in the corrosion reaction with steel. However, the city's treated supplies from surface water sources may likely contain bacteria, which may cause microbe-induced corrosion (MIC) in these stagnant, anaerobic conditions under deposits of corrosion products. It has been observed in Chicago when the wet fire protection systems are refilled that the first fill water may be turbid and foul-smelling, apparently because of contact with year-old stagnant conditions. It has also been observed that this condition is corrected by a second fill with freshly chlorinated water. Active or severe corrosion, however, has not been reported in these systems.

If there are gravity or pressurizing tanks in these systems, they should be cleaned periodically in order to minimize suspended matter, which could cause erosion of sprinkler valves.

Normally, minimum water replacement is recommended so that additional oxygen is not introduced. Another alternative is to install more corrosion-resistant materials. Galvanized steel is a somewhat better choice than plain steel, and cement-lined or plastic-lined steel, which is understood to be no more costly than plain steel, is an even better choice. Lined pipe cannot be subjected to the high heat of welding when connections are made, as the effective lining may likely be destroyed and corrosion subsequently will occur at these locations. Mechanical joints or grooved couplings can be installed to eliminate this problem. While plastic pipe would be ideal with regard to corrosion resistance, its distortion due to temperature and the release of toxic fumes on burning may be considered serious disadvantages. While standard steel or copper sprinklers are commonly specified, NFPA 13[22] tells of the installation of the new light steel and plastics being used as retrofit in sprinkler systems.

Experience with nuclear service water systems has disclosed the general superiority in corrosion resistance of copper alloys C70600 and C61400 and stainless steels 304L and 316L over cement-lined steel in closed systems. This would also apply to fire protection systems.[23]

Silicate treatment and pH adjustment, as recommended in this chapter for closed systems and in Chap. 3 for domestic hot water systems, would be a good recommendation for initial fill. This treatment would be simple to apply through the use of a shot-type feeder as the system was being filled. This treatment would also be acceptable in the potable water system so there would be no contamination of that system. Other advantages of silicate treatment are that it reduces the buildup of corrosion products and suspended matter in the system, it reduces the erosion of sprinkler valve surfaces, and it does not require supplemental treatment.

Dry fire protection systems maintained at some compressed-air pressure (such as 50 psig) would not seem to offer a corrosion problem; however, residual moisture and poor drainage caused pipe failures in one case after 8 years. One solution to such a corrosion problem would be to use nitrogen gas rather than air for pressurization, since air under pressure would be expected to increase the dissolved-oxygen content and corrosiveness of the residual water.

References

1. C. Neff, K. Smothers, M. Warnock, and M. Brooks, "Treatment of HVAC Closed Recirculating Water Systems," Report, Illinois State Water Survey, Champaign, Ill., 1988.
2. S. Sussman and H. Shuldener, "Closed Circulating Water Piping Systems," *Heating, Piping and Air Conditioning,* July 1962.
3. S. Sussman, "Is Your Closed Circulating Water System Really Closed???" *Heating, Piping and Air Conditioning,* April 1965.
4. *1991 Applications Handbook,* chap. 43: "Corrosion Control and Water Treatment," American Society of Heating, Refrigeration and Air-Conditioning Engineers, Atlanta, p. 43.1.
5. B. Kelly, *Practical Corrosion Control of Closed Hot and Chilled Water Recirculating Systems,* Nalco Reprint 134, Nalco Chemical Co., Chicago.
6. K. A. Selby, personal communication, Puckorius & Associates, 1990.

7. D. Goswami and C. N. Revellotty, "Free Cooling by Cooling Tower Water," *ASHRAE Journal,* January 1987, p. 32.
8. *Engineering and Operating Guide for Dowfrost and Dowtherm SR-1 Heat Transfer Fluids,* Dow Chemical Co., Midland, Mich., 1985.
9. M. Darrin, "Chromate Corrosion Inhibitors in Bimetallic Systems," *Industrial and Engineering Chemistry,* vol. 37, August 1945.
10. M. Darrin, "Corrosion Inhibitors in Recirculating Water Systems," *Canadian Chemistry and Process Industries,* June 1949.
11. T. Hoppe and A. McGilbra, "Hydrazine vs. Chromate as a Corrosion Inhibitor for Industrial Closed Cooling Water Systems," *Proceedings of the 38th Annual International Water Conference,* Pittsburgh, November 1977.
12. Technical Committee T7G, *Investigation of the Effects of Corrosion Inhibiting Treatments on Mechanical Seals in Recirculating Hot Water Systems,* NACE Publication TG181, National Association of Corrosion Engineers, Houston, July 1981.
13. S. Sussman, O. Nowakowski, and J. Constantino, "Experience with Sodium Nitrite, Unpredictable Corrosion Inhibitor," *Industrial and Engineering Chemistry,* vol. 51, April 1959.
14. A. Al-Borno, M. Islam, and R. Haleem, "Synergistic Effects Observed in Nitrite–Inorganic Phosphate Inhibitor Blends," *Corrosion,* December 1989, p. 990.
15. S. Shair, "Prevention of Nitrite Loss in Chilled Water Systems," *American Laboratory,* April 1983.
16. J. Conoby and T. Swain, "Nitrite as a Corrosion Inhibitor. Controlling Depletion of Sodium Nitrite," *Materials Protection,* April 1967.
17. R. W. Lane, T. E. Larson, C. H. Neff, and S. W. Schilsky, "Silicate Treatment Inhibits Corrosion of Galvanized Steel and Copper Alloys," *Materials Protection and Performance,* vol. 12, no. 4, 1973, p. 32.
18. E. Lizlovs, "Molybdates as Corrosion Inhibitors in the Presence of Chlorides," *Corrosion,* vol. 32, July 1976.
19. F. Rosa, *Water Treatment Specification Manual,* McGraw-Hill, New York, 1985.
20. A. D'ippolito, "Flushing and Chemically Cleaning New Piping Systems," *Proceedings of the 50th Annual International Water Conference,* Pittsburgh, 1989, p. 610.
21. K. D. Heinz and J. A. Gray, "Neutral pH Process for On-Line Cleaning and Passivation of Hot and Chilled Water Systems," *Proceedings of the 52nd Annual International Water Conference,* Pittsburgh, 1991, p. 383.
22. *Installation of Sprinkler Systems,* NFPA 13, National Fire Protection Association, Quincy, Mass., 1992 (phone 1-800-344-3555).
23. A. H. Tuthill, "Design, Water Factors Affect Service Water Piping Materials," *Power Engineering,* July 1990, p. 39.

Chapter

7

Water Treatment

Water quality generally refers to the many dissolved chemical species in natural groundwater, surface water, and atmospheric waters. It is the result both of the many physical and chemical weathering processes on geologic formations and of chemical reactions in the atmosphere. It is here that water lives up to its name as the universal solvent.

There is really no single definition of water quality, as there are several types, such as chemical, bacteriologic, physical, sanitary, and drinking water.[1] A specific water treatment method must usually be applied to provide the particular quality requirements of water use in the building system. This mainly pertains to potability and taste and to industrial uses requiring treatment to render the water free of scale-forming and corrosive properties.

The water source has a very definite bearing on the treatment methods needed to meet water quality requirements. While private wells may be a ready and inexpensive source, more extensive treatment equipment will likely have to be installed, and maintenance costs may become appreciable owing to the closer supervision required.

Municipal water supplies usually prove to be more reliable sources; the user is assured of having a dependable, potable, and adequately treated supply. However, the user should be aware of the source of the municipal supply and the possible variations in quality that may occur irrespective of conscientious attention to water plant operation.

In general, well waters are usually higher in hardness content [>150 mg/L (as $CaCO_3$)], are more consistent in analysis, and are more likely to be free of off-flavors and color and turbidity problems associated with some surface water supplies. Twenty or more years ago, the nation's waterways were in a sad state, but with the passage of the Clean Water Act by Congress in 1972, the amount of pollution discharged to the nation's waterways has been greatly reduced. Specifically, the U.S. Environmental Protection Agency (EPA) reported to Congress in 1992 that water pollution from "point sources," such as sewers and industrial discharge pipes carrying metal contaminants, bacteria, and oxygen-demanding organic materials, has been significantly reduced. The Clean Water Act has been revised and expanded in recent years, and industries have spent many billions of dollars to improve and enhance water quality.

Table 7.1 lists 21 common impurities found in water, the difficulties caused by their presence, and the means of treatment to correct them. References 2 and 3 provide further information on general water treatment and choices of water treatment, as well as a glossary of water treatment terms.

In practically all cases, we are fortunate to have a municipal water plant that provides us with a potable, iron-free, and stable water free of scale and corrosive tendencies at cold water temperatures. However, occasionally water plant operators may be confronted with raw-water conditions beyond their control. These conditions could be excessively turbid raw water, water main disturbances caused by fires, failure of wells producing the expected and desired water quality, and the failure of water treatment equipment and aged, corrosion-prone distribution mains. If knowledge of the local system indicates that there are frequent occurrences of these conditions, it may prove prudent to consider the installation of sand filters[2,4] to filter all incoming water.

An alternative method of solving this problem of suspended matter (down to 1 to 20 µm)[4–6] is to install in-line cartridge filters for those particular areas experiencing difficulties. The operating cost for these filters may vary from 5 to 32 cents per 1000 gal, depending on the fiber material involved, the frequency of replacement, and the flow rates involved. They may be used effectively in reducing pump impeller erosion.

Table 7.2 shows the effect of different water treatment methods on the quality of effluent water.[6,7] The choice of the treatment method[2] will depend on the uses to which the water is to be applied. For building systems, the cation exchange–sodium method will be most useful for waters exceeding 130 mg/L hardness (as $CaCO_3$). Blending with 20 to 50 percent of the hard water will then provide the desired water of 60 to 120 mg/L hardness content in the cold- and hot-water distribution systems. Water of high purity (water having a specific conductivity of 1 µS/cm) would likely only be required for desuperheating, the laboratory, or humidification. It is reported that many large buildings are not now providing humidification, but if they choose to do so, power plant steam may serve effectively.

TABLE 7.1 Common Impurities Found in Water

Constituent	Chemical formula	Difficulties	Means of treatment
Turbidity	None—expressed in analysis as units	Imparts unsightly appearance to water. Deposits in water lines, process equipment, etc. Interferes with most process uses.	Coagulation, settling, and filtration.
Hardness	Calcium and magnesium salts expressed as $CaCO_3$	Chief source of scale in heat exchange equipment, boilers, pipelines, etc. Forms curds with soap, interferes with dyeing, etc.	Softening. Demineralization. Internal boiler water treatment. Surface-active agents.
Alkalinity	Bicarbonate $(HCO_3)^-$ carbonate $(CO_3)^{2-}$ hydrate $(OH)^-$, expressed as $CaCO_3$	Foaming and carryover of solids with steam. Embrittlement of boiler steel. Bicarbonate and carbonate produce CO_2 in steam, source of corrosion in condensate lines.	Lime and lime-soda softening. Acid treatment. Hydrogen zeolite softening. Demineralization. Dealkalization by anion exchange.
Free mineral acid	H_2SO_4, HCl, etc., expressed as $CaCO_3$	Corrosion	Neutralization with alkalies.
Carbon dioxide	CO_2	Corrosion in water lines, particularly steam and condensate lines.	Aeration. Deaeration. Neutralization with alkalies.
pH	Hydrogen ion concentration defined as $pH = \log 1/(H^+)$	pH varies according to acidic or alkaline solids in water. Most natural waters have pH of 6.0–8.0.	pH can be increased by alkalies and decreased by acids.
Sulfate	$(SO_4)^{2-}$	Adds to solids content of water but in itself is not usually significant. Combines with calcium to form calcium sulfate scale.	Demineralization.*
Chloride	Cl^-	Adds to solids content and increases corrosive character of water.	Demineralization.*
Nitrate	$(NO_3)^-$	Adds to solids content, but is not usually significant industrially. High concentrations cause methemoglobinemia in infants. Useful for control of boiler metal embrittlement.	Demineralization.*
Fluoride	F^-	Cause of mottled enamel on teeth. Also used for control of dental decay. Not usually significant industrially.	Adsorption with magnesium. Used for hydroxide, calcium phosphate, or bone black. Alum coagulation.
Sodium	Na^+	Adds to solids content of water. When combined with OH^-, causes corrosion in boilers under certain conditions.	Demineralization.*

*This table was prepared before membrane processes came into general use; therefore, in each one of the above cases, the various membrane processes could likely be applicable as well as evaporation and demineralization.

SOURCE: Betz Laboratories, Trevose, Pa.

TABLE 7.1 Common Impurities Found in Water (*Continued*)

Constituent	Chemical formula	Difficulties	Means of treatment
Silica	SiO_2	Scale in boilers and cooling water systems. Insoluble turbine blade deposits due to silica vaporization.	Hot process removal with magnesium salts. Adsorption by highly basic anion exchange resins, in conjunction with demineralization.
Iron	Fe^{2+} (ferrous) Fe^{3+} (ferric)	Discolors water on precipitation. Source of deposits in water lines, boilers, etc. Interferes with dyeing, tanning, papermaking, etc.	Aeration. Coagulation and filtration. Lime softening. Cation exchange. Contact filtration. Surface-active agents for iron retention.
Manganese	Mn^{2+}	Same as for iron.	Same as for iron.
Aluminum	Al^{3+}	Usually present as a result of floc carryover from clarifier. Can cause deposits in cooling systems and contribute to complex boiler scales.	Improved filter and clarifier operation.
Oxygen	O_2	Corrosion of water lines, heat exchange equipment, boilers, return lines, etc.	Deaeration. Sodium sulfite. Corrosion inhibitors.
Hydrogen sulfide	H_2S	Cause of "rotten egg" odor. Corrosion.	Aeration. Chlorination. Highly basic anion exchange.
Ammonia	NH_3	Corrosion of copper and zinc alloys by formation of complex soluble ion.	Cation exchange with hydrogen zeolite. Chlorination. Deaeration.
Dissolved solids	None	"Dissolved solids" is measure of total amount of dissolved matter, determined by evaporation. High concentration of dissolved salts are objectionable because of process interference and as a cause of foaming in boilers.	Various softening methods, such as lime softening and cation exchange by hydrogen zeolite, will reduce dissolved solids. Demineralization.*
Suspended solids	None	"Suspended solids" is the measure of undissolved matter, determined gravimetrically. Suspended solids cause deposits in heat exchange equipment, boilers, water lines, etc.	Subsidence. Filtration, usually preceded by coagulation and settling.
Total solids	None	"Total solids" is the sum of dissolved and suspended solids, determined gravimetrically.	See "Dissolved solids" and "Suspended solids."

*This table was prepared before membrane processes came into general use; therefore, in each one of the above cases, the various membrane processes could likely be applicable as well as evaporation and demineralization.

SOURCE: Betz Laboratories, Trevose, Pa.

TABLE 7.2 Effect of Water Treatment Methods on Quality of Effluent Water

Treatment methods	Hardness, mg/L $CaCO_3$	Alkalinity, mg/L $CaCO_3$	Dissolved solids, mg/L TDS	Silica, mg/L SiO_2
Hot lime–soda	15–25	30–50	Decreased	Decreased
Hot lime–zeolite	<2	20–30	Decreased	Decreased
Cation exchange—sodium	<2	No change	No change	No change
Anion exchange—chloride	<2	15–30	No change	No change
Cation exchange—weak acid	<2	10–20	Decreased	No change
Cation exchange—split stream	<2	10–30	Decreased	No change
Demineralization	<1	<2	<5	<0.1
Reverse osmosis	90–95% rejection of the above constituents			

Clarification of a turbid or lake water may not be your concern, but in case it is, a solids-contact clarifier (Fig. 7.1) and filters, as shown in Fig. 7.2,[2] are commonly installed. Then the application of chemicals, such as Item 14 hydrated lime, Item 5 soda ash, Item 26 alum, and Item 44 chlorine[8] plus a proper coagulant will likely be required. The most efficient and proper coagulant is usually best determined through use of the laboratory "jar test" procedure at the site.[2,8–10]

The water treatment system should be designed so that the result is a "stable" water [nonscaling and noncorrosive, as indicated by the Langelier Saturation Index[7] (see Table 7.3) and described in Chap. 3 of this book]. Also, appearance (a measure of turbidity and color), taste, and odor are obvious characteristics of water that must meet the user's acceptance. Maintaining the proper hardness content, alkalinity, pH, and conductivity, which may affect the above properties and the efficiency of water use in cooling and heating operations, is important too. In order to provide a stable water free of corrosive and scaling tendencies, it is necessary to treat a surface water by means of cold lime softening. This should provide an ideal water containing 80 to 120 mg/L hardness (as $CaCO_3$), 60 to 70 percent calcium, an alkalinity of 60 to 90 mg/L (as $CaCO_3$), and a pH of 8.4 to 8.6. This quality of water should be appropriate for all general uses of water. The special uses that are the exceptions are for high-pressure boilers, 180°F (82°C) domestic hot water, and uses requiring completely softened water.

172   Chapter Seven

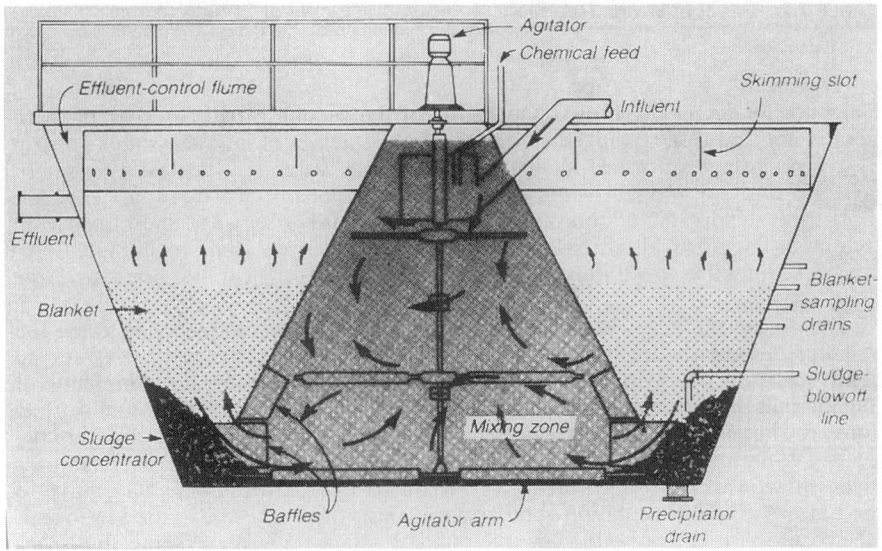

Figure 7.1 Solids-contact clarifier. [*T. C. Elliott (ed.), Standard Handbook of Powerplant Engineering, McGraw-Hill, New York, 1989.*]

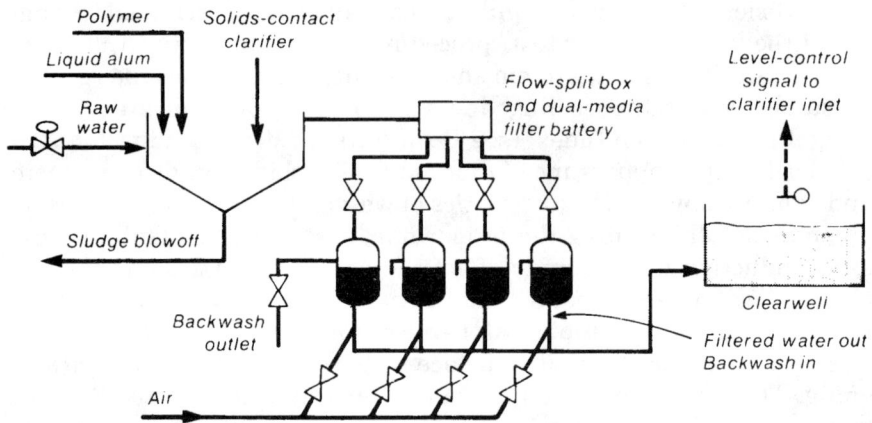

Figure 7.2 Clarifier-filter combination. [*T. C. Elliott (ed.), Standard Handbook of Powerplant Engineering, McGraw-Hill, New York, 1989.*]

TABLE 7.3 **Corrosion and Scaling Tendencies Resulting from Different Treatment Strategies**

Increased corrosion accompanied by tuberculation (pipe surface roughness and increased pumping costs)	Decreased corrosion
Negative saturation index Inadequate calcium alkalinity High Cl^-/HCO_3^- ratio High chloride and SO_4 content Chlorination Reduced alkalinity from added acidic chemicals, such as alum and fluosilicic acid	0 to positive saturation index Adequate calcium and alkalinity Low Cl^-/HCO_3^- ratio

or
Increased scale (accompanied by increased pumping costs)

High saturation index (about +0.5–1.0 or higher) caused by high calcium, alkalinity, and/or pH

Possible formation of Al_2O_3 scale by overtreatment with alum (American Water Works Association goal is to maintain Al below 0.05 mg/L in treated water)

Overtreatment with lime and lack of sufficient recarbonation to lower pH, thus causing magnesium hydroxide or silicate scale

SOURCE: Illinois State Water Survey, Champaign, Ill.

Zeolite Water Softening

Softening is not generally required for cold water used for drinking, flushing, sprinkling, etc. Completely softened water is not recommended, as it can be corrosive to metals in the system. A zeolite softener is the simplest and most inexpensive means for providing completely softened water for special needs, such as makeup for high-pressure boilers. Figure 4.3 shows a diagram of a zeolite softener, through which water is passed to exchange calcium and magnesium ions for sodium ions in order to remove the hardness from the water. The chemical reactions occurring in this exchange and in the regeneration with salt (sodium chloride) are described in Chap. 5, in the section "Prevention and Control of Scale Formation."[2]

Zeolite softener malfunctions

Occasionally, operational failures such as reduced capacity, hardness leakage, and high head losses are experienced with zeolite softeners. These failures can be caused by incorrect regeneration procedures, resin inefficiencies, or internal mechanical problems, as enumerated below:

1. *Insufficient flow or time for backwash.* At 35°F (2°C), 2.4 to 3.3 gal/(ft^2)(min) is required; at 75°F (24°C), 5.7 to 8.0 gal/(ft^2)(min) is required.
2. *Insufficient salt dosage*

Dosage	Capacity
5 lb salt/ft^3 resin	18,000 gr/ft^3 resin
10 lb salt/ft^3 resin	25,000 gr/ft^3 resin
15 lb salt/ft^3 resin	30,000 gr/ft^3 resin

3. *Weak brine solution.* It should test 26% salt, which can be readily determined through the use of a salometer (special hydrometer).
4. *Insufficient brine flow rate.* Flow rate should be 1.0 gpm/ft^3 of resin, and flow should be diluted to 10% NaCl as it contacts resin.
5. *Inadequate rinsing.* The requirement is 25 to 75 gal/ft^3 of resin at a flow rate of 1 to 1.5 gpm/ft^3; slow rinse should be for 10 to 15 min; fast rinse should be for 40 to 60 min.
6. *Internal piping.* Inspection is necessary to determine if internal piping is in need of repair. Clogged or broken distributors can cause poor regenerant distribution, channeling, and inadequate backwashing.
7. *Resin volume and condition.* An attrition loss of 1 to 3 percent a year should be expected. Also, resin can become fouled or degraded by iron deposits, biological growths, or chlorine. If it appears that the quality has deteriorated, the resin should be properly sampled and sent to the resin manufacturer for evaluation.
8. *Hardness leakage at multiport valve.* Testing for hardness leakage before and after the valve will determine whether valve and gasket repair is needed.

In summary:

Hardness should be checked periodically while the softener is in service.

Records should be kept of gallons of water softened per regeneration, salt dosage, resin volume, etc.

Brine strength should be checked weekly.

The softener should be opened and checked internally yearly.

Brine tank should be cleaned at least once a year.

POU Methods of Treatment

A method of water treatment that can be connected directly to the householder's plumbing is called a *point-of-use* (POU) method. The most commonly used POU method makes use of a cartridge filter containing activated charcoal; it may be installed to remove chlorine or objectionable tastes and odors in municipal supplies. These supplies sometimes cause an off-flavor to develop in coffee. As discussed previously, these filters may eventually serve as bacterial breeding places unless the cartridges are replaced regularly or are backwashed periodically. When these devices contain the proper chemicals, they also may be used to reduce iron and manganese fouling, high organic contents, hydrogen sulfide problems, high hardness, and corrosion and scaling problems. Their initial and continued effectiveness is based largely on the integrity of the dealer supplying and installing the equipment and on the degree of proper maintenance and service provided.

Nitrate Contamination

Nitrate, which is toxic to infants, may be particularly prevalent in shallow wells in agricultural areas, where nitrate is used for fertilizer. It may be derived from contamination of the groundwater or from municipal surface water supplies during seasons of the year when fertilizers are being applied for agricultural purposes. The maximum contaminant level (MCL) for nitrate is 10 mg/L as N (nitrogen). Nitrate can be removed from water by distillation and reverse osmosis, and now ion exchange research has brought forth a new resin that selectively removes nitrate over sulfate.[11] This anion resin, modified in structure from Type I anion resins, is regenerated with salt and has a reasonable capacity of 0.3 equiv/L (equivalent to 21.8 kgr/ft^3). POU equipment is reported to be available for home installation. The new anion resin technique is now being installed in municipal systems for use during those seasons of the year when surface water supplies may be high in nitrate.

Water Treatment Methods for Producing High-Purity Water

Table 4.2[7] shows the wide range of water quality requirements for different water uses. In general, industry is willing to accept water that meets drinking-water standards;[1] however, high-pressure boilers and certain selected industries (such as the semiconductor industry) may require the addition of extensive water treatment equipment in order to provide a required high-purity supply. The semiconductor industry requires water with a specific resistance of 15 M$\Omega \cdot$cm at 25°C (equivalent to 0.067 µS/cm conductivity). This is purer than the usual listed quality of high-purity water: resistance of 1 MΩ·cm (1 µS/cm conductivity).[7] Since water of a consistent quality is desired, pretreatment equipment is usually installed and must receive conscientious attention. Without consistent and adequate water quality, product degradation, equipment deterioration, and reduction of efficiency and capacity can be expected.

In addition to the above common water treatment methods for purifying water, there are a number of other means for producing water of high purity, such as distillation, deionization, reverse osmosis (membranes), electrodialysis, and combinations of these techniques.

Membranes (including deionization)[12]

The electric utility industry is presently installing membrane systems in combination with electrodialysis ahead of demineralizer systems, as this reduces the regenerant and disposal costs of the demineralizer process significantly. Figure 7.3 shows how membrane processes[13] differ in

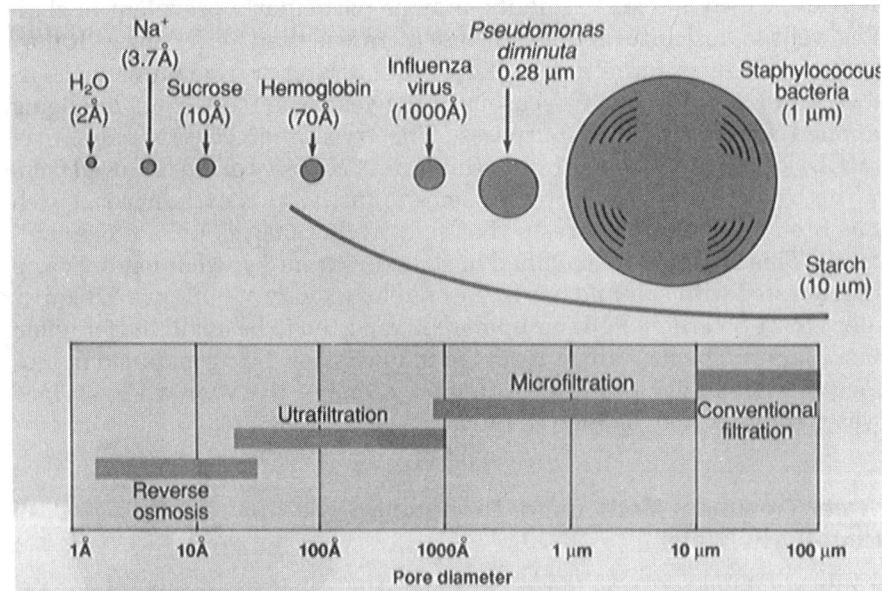

Reverse osmosis, ultrafiltration, microfiltration, and conventional filtration are related processes, differing mainly in average pore diameter of the membrane. Reverse osmosis membranes are so dense that discrete pores do not exist, and transport occurs via statistically distributed free volume areas.

Figure 7.3 Membrane processes differ in pore size. (*Chemical Engineering News*, Oct. 1, 1991.)

pore size and filtering capability (for example, the reverse osmosis, ultrafiltration, and microfiltration processes). [Å stands for angstrom units (1000 Å = 0.00001 cm), and µm stands for micrometers, formerly microns (100 µm = 0.01 cm).] Figure 7.4 shows how the reverse osmosis process works, and Fig. 7.5 shows actual modules.

In building water systems, the requirements for water quality may not actually be this high. In time, as these techniques become less ex-

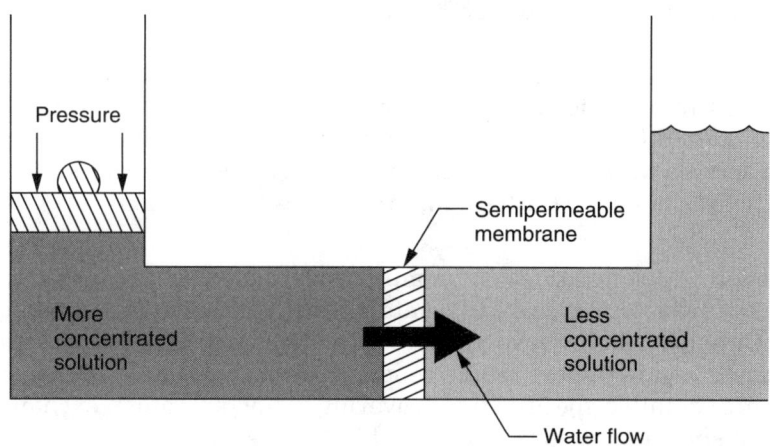

Figure 7.4 Reverse osmosis. (*Dow Chemical Co.*)

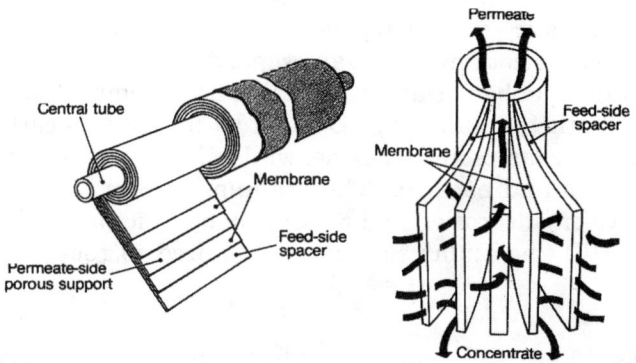

Figure 7.5 Spiral-wound, flat film module and membrane flow pattern. (*Dow Chemical Co.*)

pensive and require less skill in operation, it is to be expected that these processes will be installed in building systems. Particular advantages in heating and cooling systems will be the reduction of blowdown and the reduction of scale and corrosion problems (assuming that the proper corrosion inhibitors are added). High-purity water, however, is normally aggressive (sometimes called "hungry water") and therefore will require some treatment adjustment (such as pH adjustment). While distillation or evaporation is a simple and effective process for removing a broad range of contaminants and requires only heat and recondensing, the cost of installation and operation compared with the cost of deionization and membranes places it in a less favorable position.

In April 1992, the largest seawater desalting plant in the United States went into service at Santa Barbara, California. This system, a trailer-mounted facility capable of treating 6.7 million gal per day, relieved a serious water shortage problem and featured a partnership between the city of Santa Barbara and the Ionics Corporation.

Deionization. Deionization or demineralization[2,4,14–17] involves the removal of ions or minerals by contact with synthetic ion exchange resins. Cation resins remove positive ions such as calcium or sodium, while anion resins remove negative ions such as chloride and carbonate. Typical detailed specifications covering cation and anion exchange resins are shown as Items 48, 103, and L250 in App. A.

Removal of all ionized impurities is most effectively done by exposing water to mixed-bed deionizers, which include both cation and anion resins. Deionization can also be accomplished by separate contact with the cation resin followed by contact with the anion resin. Regeneration is necessary just as previously described for zeolite softeners, but in this case acid, such as sulfuric acid, is required for regeneration of the cation resin and caustic soda for regeneration of the anion resin system. Deionization does not remove organics, particulates, bacteria, and pyrogens (substances of bacterial origin that cause a temperature rise when injected into the human body). Contact with activated carbon is required to remove organics and chlorine, while filtration is required to remove particulates and bacteria. The granulated form of activated carbon is effective on adsorption and removal of organics and chlorine. Ultrafiltration (UF) (molecular sieves)[14] is required to remove pyrogens.

Reverse osmosis. Reverse osmosis (RO)[2,12,13] is considered a percentage removal technology because it rejects only 90 to 95 percent of the impurities found in potable water. Its usage in the preparation of high-purity water is in pretreatment, before demineralization, and the advantages are a significant savings in regenerant costs and disposal of wastes. Feedwater flows over the surface of the membrane, a percentage of the water is forced by pressure through the membrane, and a portion, high in contaminant content, is rejected and flushed to waste. Present membrane materials include cellulose acetate, polyamide, and a thin film composite, each of which possesses its own properties with respect to capacity, life, and resistance to a particular water's properties. Pretreatment is recommended, as suspended matter, chlorine, temperature, pressure, scaling tendency, iron, silica, and bacteria may affect the serviceability of the different membranes. In addition to conventional reverse osmosis, nanofiltration (NF), ultra-low-pressure RO (or membrane softening), ultrafiltration (UF), microfiltration (MF), electrodialysis (ED), and dialysis are other membrane processes that are available.[14]

Nanofiltration. The NF membrane process is a relatively new pressure-driven process. It is useful for removal of hydrated divalent ions and organic molecules with molecular weights greater than about 400 to 1000.

Ultrafiltration. The UF process, which uses a membrane having a pore size of less than 0.005 µm, is effective in removing pyrogens, colloidal silica, high-molecular-weight organics, and microorganisms. The pore size of the UF membranes is about one order of magnitude larger than the pore size of NF membranes, and UF membranes have been used in industrial water and wastewater treatment for some time.

Electrodialysis. Electrodialysis (ED) is an electrochemical separation process in which ions are transferred through membranes from a less concentrated to a more concentrated solution as a result of the flow of direct electric current. The electrodialysis reversal (EDR) process[2] is a superior process because it reverses the current flow and thereby keeps the membrane surfaces clean. This method of treatment is then able to handle waters of relatively high fouling tendencies and to achieve concentrated streams, which are supersaturated with respect to salts of limited solubility, like calcium sulfate. This capability makes the process superior to other membrane processes, demineralization, and evaporative systems with regard to capital and energy costs.

Dealkalizers

The inclusion of dealkalizing systems along with zeolite softeners is an important consideration in designing water treatment for steam boiler systems employing makeup water supplies with a total alkalinity above 50 mg/L. Dealkalizing systems appreciably reduce the amine requirements for inhibiting condensate return corrosion, which sometimes requires dosages above health limits to attain effective neutralization and corrosion inhibition. There are actually three different types of dealkalizer systems that one can consider:

1. Conventional split-stream system utilizing one acid-regenerated and one salt-regenerated polystyrene exchanger to arrive at the desired alkalinity for the effluent.
2. A salt-regenerated exchanger followed by a carboxylic acid–regenerated exchanger. This resin, although more expensive, has a high capacity and lower regenerant costs, and the system requires less rigorous control.
3. A salt-regenerated anion exchange resin following a sodium zeolite softener. This technique has the disadvantages of requiring increased boiler blowdown and higher regenerant costs, but it does not require acid—an advantage to the plants that are smaller and have lower pressures.

Iron and Manganese Removal

Soluble iron and manganese can be removed by filtration through green sand (manganese alumina silicate) and by reverse osmosis membranes, but iron must not be oxidized; otherwise, it will foul resin and membranes. Also, a zeolite softener is effective in removing iron, but if the combined iron and manganese exceeds 0.3 mg/L, periodic cleaning with sodium sulfite or bisulfite is required.

Conventional iron removal involves aeration, chlorination, and filtration. A low-cost alternative is to treat unaerated groundwater with silicate and chlorine simultaneously.[18] The process does not remove iron, but it prevents the appearance of objectionable turbidity, precipitates, and color and staining associated with iron. In stabilizing a maximum of 2.5 to 3.0 mg/L iron, 1 to 20 mg/L silicate (as SiO_2) is required and enough chlorine to oxidize the iron and provide a free residual. This technique has been successfully applied in hundreds of supplies, mostly in Ontario.

Ultraviolet radiation is used as an effective means of bacteria control and in breaking down organics for later removal by ion exchange. Ozone is being applied more and more for organics removal and as a superior bactericide for municipal potable water and waste treatment plants. One advantage is that it produces water without taste and odor problems. Its advantages in the treatment of algae and biological growths in cooling tower systems are described in Chap. 5. Recent claims for ozone in the treatment of cooling waters have included corrosion and scale inhibition; however, making use of these claims in the engineering design for new systems is still in the uncertain stage.

Gadgets

For fifty years or more, the water treatment industry and users have been plagued by promoters of water treatment gadgets, which are supposed to solve all the various water problems: scale, off-flavors, foulants, corrosion, etc. The author believes that one of the major accomplishments in supervising a state institutional water treatment program was preventing the installation of these fraudulent devices.

Over the years, the different devices have embraced so-called magnetic, electronic, electrolytic, magnetohydrodynamic, galvanic, and sonic techniques, as well as magic metals and various combinations of techniques.[19,20] The typical literature on these devices has generally claimed that passage of the water through the special fitting (perhaps resembling a swing-check-valve body) or device, as shown in Fig. 7.6, will cause the scaling and corrosive tendencies of the water to be eliminated without chemical treatment. Usually, a long discussion in pseudo-scientific jargon, such as a change in the electronic orbits or attachment of various

Figure 7.6 Gadget. (*Illinois State Water Survey, Champaign, Ill.*)

atoms, is included as a part of the promotion of the gadget. The device is sometimes offered at no cost to the customer; however, later it may prove costly, since operation without the usual and effective chemical treatment may result in outages and repair costs. In recent years, devices embracing magnetism have been particularly popular; however, research to date has not disclosed successful applications. The author's advice to those purchasing water treatment equipment is to be sure that the equipment is based on sound water treatment principles and to rely on reputable consultants for their advice.

Chemical Feed Systems

The application of the right amount of chemicals to the right location at the right time is essential in obtaining the expected corrosion and scale inhibition. In writing this book, the author has considered as his most important duty emphasizing this need to designers and building superintendents. Experience has shown that treatment control, particularly for cooling water, is most accurate when based on water meter data.[21,22] In the past, less precise treatment methods were prescribed.

In the past, accurate chemical feeding systems consisted of a makeup electricontact water meter that actuated the chemical pump at set gallonages for manually set time periods and similarly actuated a timer-set solenoid valve to provide blowdown control. Chemical pumping equipment should be wired so that the chemical pump does not operate when the air-conditioning chiller is not operating. This prevents overfeed of chemicals during outage periods.

A less expensive type of treatment system provided a conductivity instrument for blowdown control and for actuating the chemical pump for manually set time periods following actualization of the blowdown solenoid valve. Experience has shown that this type of system can lead to erratic chemical feed. For example, if losses from the system occur and conductivity tests remain low, no treatment is applied, and serious scale, corrosion, and biological problems can develop.

On the basis of knowledge of the size of the equipment, the required makeup water and blowdown may be calculated (Chap. 5) and the size of the makeup water meter determined. Then, knowing the dosage requirements of the chemicals to be applied, one may calculate the size of the chemical tank required for one week's operation. For air-conditioning systems of 500 to 1000 tons, a 50- or 100-gal plastic tank including an electric mixer usually proves adequate. If the solution strength is kept at a maximum of 10%, this means that not more than 83.5 lb of solid chemical should be dissolved in a 100-gal chemical tank.

If a blended liquid chemical (likely composed of about 15% dissolved solids) is being purchased, this means that a maximum of 40 gal should be applied to a 100-gal tank. One of the advantages of purchasing a blended liquid chemical is that the chemical may usually be applied by a chemical pump in the form received to the desired point in the water system. It has been common to purchase heavy liquid chemicals or blended liquid chemicals in 55-gal tanks; however, cleaning and disposal of these drums has become an environmental problem.

Many of the chemical companies are now providing permanent storage tanks, which they periodically fill from their tank trucks. This has relieved a serious environmental problem. An alternative for the firm purchasing generic or heavy chemicals in bulk is to install a permanent larger-sized tank and arrange for the supplier to deliver specified liquid chemicals by tank truck.

In addition to the tank and mixer, a small positive displacement chemical pump with a capacity of approximately 0.5 to 1.0 gal/h plus a microprocessor for providing the necessary communication between the water meter and the chemical pumping equipment are necessary. It is preferable that the chemical discharge point be in the cooling tower basin at an appreciable distance from the cooling tower outlet. This is particularly necessary in the case of application of acid treatment. A separate acid-resistant chemical pump, which is responsive to pH meter control, is required when acid treatment is to be applied. A plastic pan or pail should be provided at the point of acid injection in the cooling tower so that local acid concentrations will not contact metallic equipment and the solution will be diluted by overflow from the pail before contacting metals in the system. If convenient, sampling for pH, conductivity, and treatment chemicals should be done before process use; however, sampling should definitely be done after the treatment is well mixed in the system.

Present microprocessor-based chemical feed systems[22] are advantageous in that they allow easy adjustment of the feed rate, continual monitoring of water and treatment levels, and storage of chemical treatment records—all at a distance from the facilities. And all the data are easily recorded on a PC. Manual water testing is still necessary, as automated equipment still fails and isn't completely foolproof; however, manual testing may be required much less frequently.

References

1. S. D. Faust and O. M. Aly, *Chemistry of Natural Waters,* Ann Arbor Science, Ann Arbor, Mich., 1981.
2. S. D. Strauss, "Water Treatment," *Power,* June 1973, p. S-1.
3. "Glossary of Water Treatment Terms," *Water Technology,* December 1988, p. 8.
4. T. C. Elliott (ed.), *Standard Handbook of Powerplant Engineering,* chap. 4.2: "Key Treatment Systems," and chap. 4.3: "Ion Exchangers," McGraw-Hill, New York, 1989.

5. K. R. Olsen, "How to Remove Fouling Solids from Cooling Waters," *Chemical Engineering,* May 1992, p. 155.
6. B. S. Parekh, "Get Your Process Water to Come Clean," *Chemical Engineering,* January 1991.
7. J. M. Montgomery Engineers, Inc., *Water Treatment Principles and Design,* John Wiley & Sons, New York, 1985, p. 314.
8. G. C. White, *The Handbook of Chlorination,* Van Nostrand Reinhold, New York, 1986.
9. H. E. Hudson, Jr., *Water Clarification Processes,* Van Nostrand Reinhold, New York, 1981.
10. S. K. Dentel, B. M. Gucciardi, T. A. Bober, and J. J. Resta, *Procedures Manual for Polymer Selection in Water Treatment Plants,* AWWA Research Foundation, Denver, 1989.
11. L. S. Golden and J. Irving, "Development of a Resin to Selectively Remove Nitrates from a Water Supply," *Proceedings of the 51st Annual International Water Conference,* IWC Paper 90-17, Pittsburgh, 1990, p. 213.
12. J. Haggin, "Membrane Technology Has Achieved Success, yet Lags Potential," *Chemical & Engineering News,* Oct. 1, 1990, p. 22.
13. T. C. Blanton, D. Rohe, J. C. Jacangelo, and B. J. Marinas, "Emerging Membrane Processes for Drinking Water Treatment," *Waterworld News,* January–February 1991, p. 10.
14. *The Barnstead Basic Book on Water,* 1971, and *The Water Book,* 1991, Barnstead Co., Boston.
15. G. P. Simon, *Ion Exchange Training Manual,* Van Nostrand Reinhold, New York, 1991.
16. D. L. Owens, *Practical Principles of Ion Exchange Water Treatment,* Tall Oaks Publishing Co., Voorhees, N.J., 1985.
17. *Engineering Manual for the Amberlite Ion Exchange Resins,* Rohm & Haas Co., Philadelphia, 1979.
18. R. B. Robinson, R. A. Minear, and J. M. Holden, "Effects of Several Ions on Iron Treatment by Sodium Silicate and Hypochlorite," *Journal of the American Water Works Association,* July 1987, p. 116.
19. R. M. Westcott, "Nonchemical Water Treating Devices," *Materials Performance,* November 1980, p. 40.
20. F. Rosa, *Water Treatment Specification Manual,* McGraw-Hill, New York, 1985, p. 190.
21. Nalco Chemical Co., *The NALCO Water Handbook,* chap. 45: "Chemical Feed Systems," McGraw-Hill, New York, 1988.
22. D. Morris and J. Kuchinski, "Microprocessor Based Cooling Tower Control—An Overview," *Proceedings of the 51st Annual International Water Conference,* Pittsburgh, 1990.

Chapter

8

Materials

The selection of materials is an essential item in designing corrosion-free systems. Different materials have properties of resisting deterioration under different environmental conditions and different cost factors.[1] Since hardness scale forms on all the commonly used materials, one need not consider this factor in choosing a material.

On the basis of the National Plumbing Code of the Building Officials and Code Administrator (BOCA),[2] water service piping materials may be asbestos cement, brass, cast iron, copper, plastic, galvanized iron, and steel. Distribution water piping may be brass, copper, plastic, and galvanized steel. Local restrictions may limit the choice of these materials.

Joints between pipe and fittings must be firm, and adequate hangers and supports must be provided to ensure piping integrity. When plastic pipe is installed, particular attention must be paid to the materials and quality of work involved in the joining of piping and fittings. Also, adequate supports must be installed for this piping, which is light in weight and which expands appreciably with temperature change because of its high coefficient of expansion.[3]

Since most facilities will be supplied with water from a municipal water plant, the details of designing a system for a private water supply are not included in present discussions. For those interested, reference should be made to Chap. 3 of Ref. 3, which provides more details on such installations.

Materials to be installed in a new facility should be evaluated beforehand by obtaining information from a qualified corrosion engineer familiar with the project and environmental factors involved and by contacting building managers at nearby structures. When one is completely assured that the same water conditions exist [for example, temperature, pressure, water analysis (including gases), contact with other metals, velocity, metallic surface conditions, water treatment, and potentiality for the development of crevice corrosion and microbe-induced corrosion (MIC)], one can be assured that similar materials and water treatment should prove satisfactory. In areas where information on past experience is unavailable and initial investigation discloses that serious corrosion is a likely possibility, laboratory testing of materials under the varied expected environmental conditions should be arranged.[4-6]

The corrosion of the commonly installed materials, namely copper and galvanized steel, is covered in Chap. 2, and so now the corrosive tendencies and service life of other materials, including stainless steel, copper-bearing alloys, and plastic piping, will be discussed.

Cast iron, in many cases lined with coal tar or cement, has been used successfully at cold water temperatures in municipal distribution systems for many years. Although cast iron is inexpensive and mainly corrosion-resistant owing to its thickness, iron corrosion products formed in the piping, particularly in aggressive waters and at dead ends, may detract from the water quality derived from it. It is essential that the water in contact with cast-iron piping be stable [properly treated to maintain a Langelier Saturation Index (LSI) of 0.0 or above]. Cast iron is now generally being replaced by ductile iron, which is somewhat more corrosion-resistant.

Steel is less corrosion-resistant than cast iron, and therefore its usage is confined mainly to the boiler room, where controlled corrosion-inhibited waters are in use. Experience has revealed that steel is an unsatisfactory material for domestic hot water lines, even when the water is properly treated with a prescribed sodium silicate.

Galvanized steel, discussed in detail in Chap. 3, suffers serious corrosion in low-hardness waters and is suitable mainly for the harder, more alkaline waters of the midwest. Even in these cases, silicate or other inhibitor treatments may be required in the more aggressive waters.

Quality of Work

In some cases, the failure of galvanized piping at fittings may be the result of poor quality of work in threading pipe; there may be increased exposure of the steel under the galvanized coating, resulting in increased corrosion of the exposed steel. Abiding by recommended threading depth and length and using Teflon tape or pipe "dope" ensure minimum steel exposure and subsequent corrosion. These precautions can be particularly effective in the case of galvanized pipe near brass valves.

In Ref. 7, a study of how quality of work influenced the dissolution of tin-lead solder used in joining copper tubing revealed that this was an important factor. In other words, more solder surface was exposed by some workers than others, and as a result, a higher lead content was found in potable water where there was a greater exposed solder surface. The use of non-lead-containing solders has now corrected this problem in all new construction.

Copper is the preferred piping material; however, the piping system must be designed so that velocities do not exceed 4 ft/s (1 m/s) and turbulence is avoided. In recent years, it has been disclosed that the tin-lead solder used in joining copper piping has caused undesired lead contamination of potable water; it is now required that this solder be replaced by tin-antimony and tin-silver solders. Higher tolerance to velocity and turbulence is attained by substituting Admiralty metal and the cupronickels (90-10 and 70-30) for copper.

Reference should be made to Chap. 5 to obtain information on the corrosion resistance of materials exposed to cooling waters, particularly in cooling towers.

Stainless Steels

The stainless steels are more expensive and therefore impractical for some uses, but their advantages are that they do not contribute corrosion products to potable water and are not limited in application at the higher velocities.

Stainless steels serve as suitable materials for chilled drinking water systems to avoid turbidity, off-flavors, and corrosion products. Galvanized iron and steel used for this service have been known to experience corrosion and cause the formation of corrosion products that are detrimental to the desired water quality.

While there are some 57 stainless steels recognized by the American Iron and Steel Institute as standard alloys, our concern in freshwater systems is mainly with the austenitic stainless steels: 304 (frequently referred to as 18-8 stainless), 304L, 316, and 316L. The 316 stainless steel is more corrosion-resistant than 304. The 304L and 316L, containing less carbon, can be heated in the sensitizing temperature range (as in welding) and are not susceptible to intergranular corrosion.[6]

Like other materials, stainless steel has its deficiencies. It is necessary to keep its surface clean to avoid discoloration, pitting, crevice corrosion, and MIC. Also, one should be aware of its property of developing stress corrosion cracking in chloride environments. Since the corrosion resistance of stainless steel is dependent on maintaining a thin surface oxide film, it is understandable that its surface must be kept clean to maintain ready contact with oxygen from the air. As might be expected, stainless steel is subject to corrosion under crud (crevice corrosion). It is also subject to MIC when low- or stagnant-flow conditions are allowed to persist. In fact, a flow rate of 10 ft/s is recommended in the design of water systems containing stainless-steel piping.[8] The question arises as to how low a flow rate should be avoided. It is estimated that 1 ft/s is too low and that 4 ft/s may be adequate to keep suspended matter from settling out and causing crud buildups and areas of differential aeration.

In the case of stress corrosion cracking, keeping temperatures below 140°F (60°C) even in the presence of high chlorides and stress usually provides the necessary inhibiting effect. Of course, there are other stainless steels that can be used, such as the ferritic stainless steels (405 and 430), which are resistant to stress corrosion cracking (SCC).[8]

In comparing stainless-steel pipe with fiberglass pipe, one sees a significant economic advantage to installing fiberglass pipe; however, the advantage may vary, depending on the pipe diameter. Cases of pipe failure had indicated a distinct advantage for stainless steel, but more recently, the installation of more adequate pipe supports for the fiberglass piping has apparently corrected these cases of failure.[9,10]

Copper-Bearing Alloys

In processes in which temperatures are maintained above 140°F (60°C) and the velocities are above 4 ft/s, galvanized steel and copper may experience some corrosion. An example would be completely softened water piped ahead of dishwashers or laundry operations. In these cases, stainless steel or 90-10 cupronickel alloys containing a small amount of iron have proved corrosion-resistant and especially resistant to the higher temperature, turbulence, and high velocity.[11]

Protective Coatings[12]

Protective coatings are another important means of inhibiting corrosion, in addition to making the correct choice of materials and applying corrosion-inhibiting chemicals. While there are many types of protective coatings, probably the most commonly used and adaptable are organic coatings (paint films), which will now be discussed. These coatings, particularly important for carbon steels, serve as an insulation from the corrosive electrolyte and therefore disrupt the corrosion process.

Proper surface preparation is most important in attaining an effective protective coating. This surface preparation is based on specifications prepared by the Steel Structures Painting Council (SSPC).[12] Without proper surface preparation, the best protective coating can fail. Methods for preparing the surface may involve abrasive cleaning, steam and water cleaning, flame cleaning, pickling, solvent or vapor degreasing, or wire brushing. Abrasive cleaning in the form of sandblasting or gritblasting is favored.

Protective coatings are usually applied in three steps: a prime coat to bond the intermediate and top coats to the metal surface, followed by the intermediate and top coats. There are twenty or more different organic coating systems available that are suitable for a particular purpose and environment. The frequently used classes of coatings are epoxies, urethanes, vinyls, chlorinated rubbers, phenolics, acrylics, coal tar, zinc rich, alkyds, and combinations of the above.

Table 6.2 in Ref. 13 lists the generic names, methods of application, and adaptability to different environments. Reference to the *NACE Coatings and Linings Handbook*[14] is helpful in selecting the proper coating. Steel specimens with a protective coating are supplied by the coating manufacturers so that the customer can test the coating for its corrosion-inhibiting ability. After a coating is applied, a qualified inspector can examine the film and measure its thickness.

Plastic Piping[15,16]

Plastic piping is an important choice of material to limit corrosion in a corrosive environment, and it is becoming more acceptable as a substitute for metals. Its physical properties have been improved, and its limitations are now better recognized and understood.

Plastic piping includes both thermoplastics and thermosets.

Thermoplastic piping is mainly extruded, although fittings are injection-molded (fabricated from extruded pipe). Thermosets, or reinforced thermosetting resin piping and fittings, are made by four different processes: filament winding, centrifugal casting, contact molding, and compression molding.

Thermoplastic materials have several advantages: they are resistant to corrosion and deposition; they are light in weight; they are economical; and they have a lower modulus of elasticity and a lower coefficient of friction than metal piping. However, these materials have been shown to possess the following deficiencies: (1) permeation by organic solvents (such as petroleum products) and (2) a high thermal coefficient of expansion.

Numerous cases of the permeation of underground plastic piping by petroleum products or other organic solvents[17,18] have caused serious contamination, off-flavors, and odors in potable water. Its high thermal coefficient of expansion makes it necessary to install numerous expansion joints, and as a result, there is a greater likelihood that leaks will develop. The lesser mechanical strength, the need for more frequent supports, and the joining of fittings may complicate the installation and lead to a greater leaking problem.[18]

There are six different plastic piping compositions[15] being used in potable water systems:

1. *PVC (polyvinyl chloride): Type 1, Grade 1.* This plastic piping is most frequently specified and has been used for more than 30 years in chemical processing, industrial plating, chilled distribution systems, deionized-water piping, chemical drainage, and irrigation piping. It is resistant to acids, alkalies, and salts but is attacked by polar solvents, such as ketones, chlorinated hydrocarbons, and aromatics. Its maximum service life is attained below 140°F (60°C), but it has the highest long-term hydrostatic strength at 73°F (23°C) (design stress of 2000 psi) of the major thermoplastics. It is joined by solvent cementing, threading, or flanging.

2. *CPVC (chlorinated polyvinyl chloride): Type 4, Grade 1.* Its physical properties are similar to those of PVC, its chemical resistance properties are somewhat better. It also possesses design stress properties to 2000 psi. Its maximum service life is attained below a maximum temperature of 210°F (99°C), and it has been used successfully for hot and cold water distribution for 30 years. It is joined by the same procedures as for PVC and can be installed at substantial labor savings over the installation costs for metal pipe.

3. *PP (polypropylene): Type 1.* This is a lightweight polyolefin. It is generally high in chemical resistance to acids, alkalies, and organic solvents but is at a somewhat lower level in physical properties. It can't be used with strong oxidizing acids, chlorinated solvents, and aromatics. It has a design stress of 1000 psi at 73°F (23°C). It is resistant to sulfur-bearing water in saltwater disposal lines and in piping carrying crude oil and low-pressure gas. It is joined by thermoseal fusion, threading, and flanging.

4. *ABS (acrylonitrile-butadiene-styrene): Type 1, Grade 3.* ABS is a combination or blend of polymers in which minimum butadiene is 6 percent, minimum acrylonitrile is 15 percent, minimum styrene is 15 percent, and all other monomers are not more than 5 percent; ABS also contains other additives. It is an excellent resin for piping, because of its toughness, strength, and stiffness. These properties account for its extensive use in drain, waste, vent, sewer, and communications ductwork.

5. *PB (polybutylene)*. PB is produced by the polymerization of butylene and normally contains 2 percent carbon black as an ultraviolet inhibitor. It is well suited for piping because of these properties: it is flexible with long-term strength; it is noted for retaining tensile strength better at temperatures from −10 to 200°F (−23 to 94°C) in pressurized systems; and it is abrasion-resistant.
6. *PE (polyethylene)*. This is the second most widely used thermoplastic material. Normal additives are antioxidants and carbon black to screen out ultraviolet light. There are three types: Type 1 has low density, is relatively soft and flexible, and has low heat resistance; Type II has medium density, is slightly harder, and is more heat-resistant and higher in tensile strength; Type III is the preferred piping material owing to its toughness and superior physical properties.

Plastic piping (polybutylene) has been installed in residential subdivisions in which serious leakage problems developed and resulted in extensive liability suits. However, such incidences are rare and are not usually the result of an inferior material; more often they are the result of inferior quality of work or of not using proper joining procedures.

According to *Review of Water Industry Plastic Pipe Practices,* published by the American Water Works Association Research Foundation,[19] there is a greater satisfaction with metal services than with any type of plastic service. Also, the major problem associated with plastic services is in the installation (for example, joining of fittings). However, it is reported that there is a high degree of satisfaction with the installation of PVC mains. Also, it is a rather common practice to install polyethylene encasements to provide corrosion protection for new cast-iron and ductile-iron water mains.

While most plastic pipe is limited to pressures below 150 psi and temperatures below 200 to 300°F (94 to 149°C), it is estimated that it could be used to the extent of 60 to 80 percent of industrial piping applications rather than the present 5 to 10 percent.[20] Some of these uses would be for domestic water, air and natural gas distribution, compressed air, and underground sewers and utilities.[21] As plastic pipe is a poor heat conductor, it should be recognized that insulation costs can be reduced or eliminated.

Besides the thermoplastics, there are fiber-reinforced plastics (thermosets), which may be reinforced with silica, fiberglass, or graphite. Probably the one best known and most used is FRP, which is reinforced with fiberglass.

Lined pipe—metal pipe with a plastic liner—is another corrosion-resistant alternative. It has the advantages of chemical resistance and the structural strength and shock resistance of the metal outer shell. Flanged connections are necessary with lined pipe; also, welding must be barred, and costs may be higher.[22]

Corrosion of Valves

Since it is necessary that valve seats retain dimensional accuracy and be free of pitting, the trim shall be designed to be cathodic to the valve body, and consequently, the valve body should be steel and not cast iron. Graphitic corrosion of the cast-iron body can lead to the development of graphitic cathodes and subsequent corrosion of the trim or valve seats.

Streamlining the equipment by eliminating as many bends, valves, branches, and flow controllers as possible in order to reduce entrained air and turbulence will result in reduced erosion-corrosion. The installation of diaphragm valves rather than globe or gate valves will also cause less turbulence. Plastic-lined valves should be installed in cases in which corrosion is expected under turbulent conditions.

The installation of proper valves for the service involved is an important item in valve maintenance.[23] Gate valves should be used for providing on-off conditions but not for throttling; globe (and angle, needle, and Y-pattern) valves are designed for throttling. Corrosion of silicon red brass valve stems[24] has been reported; in these cases, dezincification has occurred but is now controlled by suitable additives, such as arsenic and phosphorus. Suspended matter can be erosive to valve seats, and so if significant suspended matter (for example, corrosion products) is present in the flow, consideration should be given to the installation of a small in-line filter for removing the suspended matter.

When the optimum materials of construction for valves are planned, the valve body and trim (including seats) need to be considered separately because of the difference in liquid velocity occurring at these locations. The installation of extremely hard materials at seats or plugs can usually be justified. For low-temperature, low-pressure water service, iron and either bronze or plastic are the usual recommended materials, and the stainless steels are recommended for extreme pressure and temperature conditions. Since new materials, particularly plastics, may be applicable, the supplier should be contacted to obtain a recommendation for a valve for a particular service.

The chapter entitled "Materials for Particular Applications in Water Systems" in Ref. 11 lists the characteristics of materials for water piping systems. In general, agreement is shown with information already provided in this book. Reference to this book is recommended to obtain information about the expected service of different materials in valves, pumps, condensers, and heat exchangers.

Corrosion of Pumps

Pumps present a particular corrosion problem because of the high velocity and turbulence involved. Cavitation, a form of impingement corrosion, is observed in pump impellers and is the result of the collapse of air or vapor bubbles on the metal surface, as a result of the repetitive development of low- and high-pressure areas. Figure 8.1 is a good illustration of the effect of corrosion on a bronze pump impeller exposed to a midwestern water for a few months in circulating domestic hot water service. Replacement of the bronze impeller with one made of steel, a much harder metal, corrected this erosion-corrosion problem. While high velocity generally causes increased corrosion, minimum velocity or stagnation, as experienced with plant shutdowns, is also recognized as a serious cause of increased corrosion.

The liquid end of centrifugal pumps is generally provided in bronze-fitted, all-bronze, or iron-filled construction. The impeller, shaft sleeve,

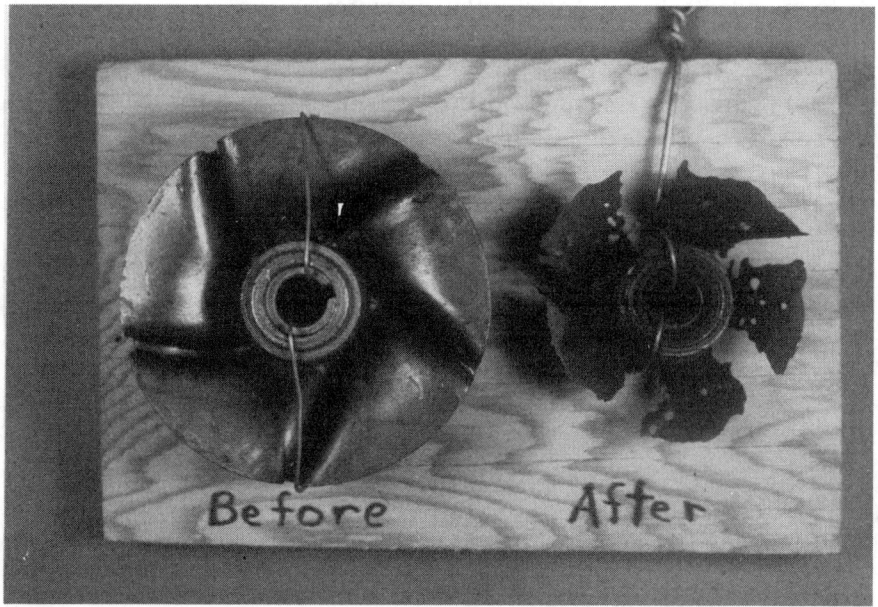

Figure 8.1 Erosion-corrosion of bronze pump impeller. (*R. W. Lane files.*)

and wear rings are bronze and the casing cast iron in bronze-fitted construction. The corrosion-resistant plastics, such as Teflon, Hypalon, Neoprene, Buna N, Viton, EPDM, Kynar, Noryl, and Ryton, are now being used for major and minor component parts of pumps. If adequate attention is paid to the material's properties of mechanical strength and thermal resistance and to the pump manufacturer's advice, purchased equipment should be of the desired efficiency and should have a long life.[25]

Packing or a mechanical seal is provided in the stuffing box, where the rotating shaft enters the pump casing. Mechanical seals are used mostly in hydronic applications.

Pump Mechanical Seals

In 1971, Committee T7A of the National Association of Corrosion Engineers (NACE) investigated the reasons for premature failure of single-face mechanical seals and concluded that the major causes were as follows:

1. Improper installation of seal or pump
2. Dry or partially dry operation
3. Suspended solids in the circulating water
4. Excessively high water temperature

After this investigation, opinions were expressed that the role of water treatment chemicals and their treatment levels should be investigated; accordingly, NACE Committee T7G1 undertook this investigation. A report entitled "Investigation of the Effects of Corrosion Inhibiting Treatments on Mechanical Seals in Recirculating Hot Water Systems" was published in 1981.[26] It disclosed the following:

1. Chromate treatment above 500 mg/L (as CrO_4) causes an increase in carbon seal wear.
2. Nitrite treatment up to 4000 mg/L (as NO_2) causes no increase.

References

1. B. Waller, "Piping—from the Beginning," *Heating/Piping/Air Conditioning*, October 1990, p. 51.
2. Building Officials and Code Administrator's National Plumbing Code, BOCA, Country Club Hills, Ill., 1987.
3. J. F. Mueller, *Plumbing Design and Installation Details*, McGraw-Hill, New York, 1987.

Materials 195

4. G. N. Kirby, "How to Select Materials," *Chemical Engineering*, Nov. 3, 1980, p. 86.
5. R. J. Landrum, *Fundamentals of Designing for Corrosion Control*, National Association of Corrosion Engineers, Houston, 1989.
6. A. H. Tuthill, "Design, Water Factors, Affect Service Water Piping Materials," *Power Engineering*, July 1990, p. 39.
7. AWWA Research Foundation, *Lead Control Strategies*, American Water Works Association, Denver, 1990.
8. R. S. Brown, "Selecting Stainless Steel for Pumps, Valves, Fittings," *Chemical Engineering*, Mar. 9, 1981, p. 109.
9. Committee of Stainless Steel Producers, *Design Guidelines for the Selection and Use of Stainless Steel*, American Iron and Steel Institute, Washington, D.C., 1977.
10. F. Britt, "Stainless Steel vs. Fiberglass Pipe," *Chemical Engineering*, February 1990, p. 105.
11. K. R. Tretheway and J. Chamberlain, *Corrosion for Students of Science and Engineering*, J. Wiley & Sons, New York, 1988.
12. *Surface Preparation Specifications*, ANSI A159.1-1972, Steel Structures Painting Council, Pittsburgh, 1972.
13. *A Practical Guide to Coating Work in Power Plants*, Steel Structures Painting Council, Pittsburgh, 1991.
14. *NACE Coatings and Linings Handbook* (loose-leaf), National Association of Corrosion Engineers, Houston, 1985.
15. D. A. Chasis, *Plastic Piping Systems*, Industrial Press, Inc., New York, 1988.
16. Nibco, Chemtrol, *Plastic Piping Handbook*, Indianapolis, 1989.
17. J. K. Park, L. Bontaux, T. M. Holsen, D. Jenkins, and R. Selleck, "Permeation of Polybutylene Pipe and Gasket Material by Organic Chemicals," *Journal of the American Water Works Association*, October 1991, p. 71.
18. T. M. Holsen, J. K. Park, D. Jenkins, and R. E. Selleck, "Contamination of Potable Water by Permeation of Plastic Pipe," *Journal of the American Water Works Association*, August 1991, p. 53.
19. AWWA Research Foundation, *Review of Water Industry Plastic Pipe Practices*, Research Report, American Water Works Association, 1987.
20. J. A. Bachetti, "Plastic Pipe Can Cut Maintenance Costs," *Maintenance Technology*, November 1989, p. 33.
21. T. Sixsmith, "How to Select the Right Plastic Pipe," *Plant Engineering*, Dec. 15, 1988, p. 40.
22. W. O'Keefe, "Special Report on Corrosion-Resistant Piping," *Power*, April 1981, p. S-1.
23. R. C. Merrick, "A Guide to Selecting Manual Valves," *Chemical Engineering*, Sept. 1, 1986.
24. L. P. Costas, "Field Testing of Valve Stem Brasses for Potable Water Service," *Materials Performance*, August 1977, p. 9.
25. J. L. Foszcz, "Pumps for Corrosion Resistance," *Plant Engineering*, Apr. 26, 1990, p. 49.
26. NACE Committee T7G1, "Investigation of the Effects of Corrosion Inhibiting Treatments on Mechanical Seals in Recirculating Hot Water Systems," Publication 7A170, National Association of Corrosion Engineers, Houston, 1981, reprinted in *Materials Performance*, July 1981, p. 53.

Chapter

9

Building Systems and Maintenance

The modern city office building may be 50 or more stories high and have separate zones every 7 to 10 stories. It may be considered a highrise building when the city water pressure is not high enough to reach the top of the building. A small lead pump may provide enough water flow for average usage, but supplementary pumps will likely be required to satisfy peak demands. Within each zone one may expect 50-psi water pressure at the top floor and 80 to 85 psi at the bottom floor, and pressure-reducing valves (PRVs) may likely be installed to prevent excessive pressures, which can cause valve erosion. The alternative is to install water storage tanks in each zone and thus save some pumping energy; however, such tanks occupy space, which has commercial value. The difference in pressure has minimal importance with respect to corrosion, since the solubility of oxygen in the water and the corrosion rate dependent on oxygen content are not appreciably affected.[1,2]

Domestic Hot Water

Domestic hot water is normally heated in each zone. Retaining heat is necessary in these systems; otherwise, waste occurs when tenants must draw off substantial volumes of cold water from long lines before obtaining hot water. It has been common to install a circulating line and pump so that the user has only a short distance from which to draw hot water. Other methods that are now being used include installing heaters at the point of use (POU heaters) and installing electric resistance heat tape around the hot water lines. Since hundreds of POU heaters would be required in a large building, this idea may not be considered economical, while the heat tape idea may even be usable to supplement a single hot water system in a large building. Individual engineering studies may be required to decide which method is economical for a particular application.

City codes may require that domestic hot water heaters have double-shell heat exchangers in order to provide double protection in case of corrosion penetration by the heating medium into the potable water. In addition, city-code-approved lining of steel tanks may be specified. The author's experience has indicated that hot water heaters or storage tanks should be cement-lined (see Chap. 4) or glass-lined in order to avoid corrosion of steel tanks and contamination of the supply by corrosion products.

The following factors need to be considered in deciding on the temperature to be maintained in a domestic hot water system:

1. Cost of maintaining desired temperature
2. Increased corrosion rate at higher temperature and reversal of potential of some metals at higher temperatures
3. Safety
4. Bacteriologic growth at lower temperatures
5. Volume of water required at the different temperatures
6. The intended use and temperature required

Naturally, the higher the temperature to be provided, the higher the fuel cost. While in general, the rate of a reaction doubles for each 10°C temperature increase, this doesn't always hold true in specific corrosion reactions. In the writer's opinion, a domestic hot water temperature of 135°F (57°C) is best. This, of course, must be qualified for hospitals, where patients could be burned by a temperature this high, and for dishwasher use, where 180°F (82°C) water is required for sterilizing dishes. Temperatures of 105 to 120°F (41 to 49°C) are the normal temperatures provided for direct lavatory use, but these temperatures are too low to prevent *Legionella* and other bacteria from growing. Temperatures above 140°F (60°C) can cause serious corrosion of galvanized steel and even a reversal of potential, which can cause steel rather than zinc to be anodic.[3]

Materials in Water-Using Systems

Chapters 2, 8, and others should provide important background information for making the proper choice of materials. However, one should make use of all information available in planning a new facility, and so the manufacturers of piping and condenser tube materials, valves, and pumps should be contacted to get their recommendations as to the best materials to install to attain long-life, corrosion-free, and economical operation. In choosing a material from the common metal piping materials (cast iron, steel, galvanized steel, copper, Admiralty metal, the cupronickels, and stainless steel), consideration must be given to the following:

1. The water quality and the estimated corrosion resistance of the material to this particular water quality
2. The service to which the water will be subjected
3. The past history of the material with respect to the particular water supply (if available)[3]

The stainless steels, discussed in Chap. 8, are more expensive and therefore impractical for some uses, but their advantages are that they do not contribute corrosion products to potable water and are not limited in application at the higher velocities.

Monitoring

Computer programs entitled COOLAID, HOTCALC, and COMTECH,[4] available from EPRI through the local electric utility, can be helpful as tools for commercial building energy analyses. Energy management systems involving heating, cooling, and ventilating control by computers are commonly installed in the modern office building or condominium. Since this monitoring equipment displays temperatures, pressures, humidity, flows, equipment operation, etc., it should also be used to display water treatment variables. Proper water treatment results depend very much on proper water test results day after day, and since there may be a multitude of water-monitoring locations throughout a facility, the monitoring equipment can be most beneficial in attaining the desired operating efficiency of water treatment equipment and in conserving labor. This brings up another important phase of building maintenance, namely the training and aptitude of building operators. Building operators must now work with more sophisticated equipment, such as computers, and therefore must be better technically trained than the custodians of the past. Their responsibilities are now more serious ones, and the success of the building operation may depend entirely on their conscientiousness, technical knowledge, and manipulation of the operating system.

Present monitoring of closed and open water system variables, such as pH and conductivity, can be provided from continuous measurement with in-line sensors and a data display on a PC. In this way, the operator can be made aware of these variables hourly, just the same as with airflow, temperatures, and pressures, and corrections can be made before serious corrosion or scale consequences occur. As practically all treatments that are applied alter the pH of the water system, the pH measurement provides a rough indication of whether the treatment is being applied, and the conductivity measurement indicates whether proper blowdown is being administered. Also, the turning on and off of the chemical pump or other details can be transmitted. This information can be transmitted by fax to remote locations for interpretation and monitoring. Corrosion probes can also be monitored; however, their measurement values and the interpretation of the values may be less exact and more controversial. As a recent report[5] discloses that the recent problem with "white rust" on galvanized towers may likely be the result of a difference in the aluminum and lead content of galvanizing, the need for accurate monitoring of pH by in-line measurement becomes particularly important.

One important aspect of such control is that the sensors will still require periodic checking and cleaning. The need for conscientious attention to these matters can't be overstressed, as habitual neglect of this maintenance may show up in serious corrosion failures.

Manual water testing to determine that the specified treatment levels, pH, and conductivity are being maintained properly is still required; however, such testing will need to be done less frequently. One cannot be completely dependent on automatic testing, as automatic equipment is known to occasionally malfunction; the result is serious scale and corrosion problems.

Survey of High-Rise Buildings

A survey of five high-rise (40-story) office buildings, ranging in age from 6 to 70 years, in the Chicago area has revealed a decided difference in materials and in heating and air-conditioning design. The older buildings were built with schedule 80 steel piping and have experienced a minimum of piping failures. Chicago water derived from Lake Michigan is a well-treated, medium-hard water [hardness content is 130 mg/L (as $CaCO_3$)]; it also contains appreciable alkalinity and minimum chloride and sulfate, and it is treated to a proper pH to provide a Langelier Saturation Index of 0.0 to +0.2 for control of corrosion and scale formation. Galvanized piping exposed to cold and recommended hot water temperatures of a maximum of 135°F (57°C) has had a normal life expectancy of 50 years or more in this supply.

The older buildings did not have separate systems for heating and cooling, and so it was necessary to change from cooling or heating as the seasons changed. None of these buildings in general provided humidity control,[6] for it was found that condensation was not a problem if circulating water in induction coils was held at a minimum temperature of 59 to 60°F (about 15°C). When humidity control was provided, it was accomplished by the dehumidification occurring in contact with the chilled water coils.

In some of these buildings, in which there is significant heat from building lighting, building occupants, and other sources, there was no need for operating the boilers even in the coldest winter, and so it was necessary only to regulate their chillers and circulating water temperature to maintain required circulating water temperatures. The present trend is to install electric boilers for producing steam or hot water rather than fossil-fuel-fired boilers.

Buildings near the Chicago River use this water for once-through cooling of their chillers and also make use of this cool water and heat exchangers for "free, or ambient, cooling" when cooling temperature requirements are less severe. As the river has become infested recently with zebra mussels, it has been necessary to free condenser surfaces of these growths by increasing condenser water temperatures to above 90°F (32°C) periodically. This has solved the problem fairly satisfactorily for office buildings not maintaining 24-hour operation, but those having to maintain cooled office temperatures for 24 hours are now recognizing the need for applying chemical treatment. The U.S. Environmental Protection Agency (EPA) is now requiring that National Pollution Discharge Elimination System (NPDES) permits be obtained for either thermal or chemical treatments for these buildings.

The newer buildings are using the cooling tower water as makeup for the building chilled distribution water, and as might be expected, suspended solids (airborne dirt) and the corrosive character of this water of higher dissolved solids content are causing some deposition and corrosion. In-line filters including automatic backwash are usually installed, but these are reported to cause excessive loss of treated water in some cases. Water treatment must be maintained at a somewhat higher level in order to attain corrosion inhibition, and effective water treatment dispersants are required for dispersion of the suspended matter in the closed chilled water system (see Chap. 6).

Water treatment

All parties contacted recognized the value of having their own personnel perform the water testing and control. In this way, their own personnel recognize the need for careful control to prevent their having to do extra work to correct a bad scale or corrosion problem. It is expected that the water treatment company will also test and take samples once or twice a month and check the plant testing methods.

The decision on which water treatment company to hire is based on both price and the degree of helpful service provided. The periodic installation of corrosion coupons and reports of the degree of corrosion and scale are usually included in the contract with the water treatment company.

Drinking water systems

The older office buildings may provide a 15,000-gal (or larger) water tank on the top floor and will have this supply tested periodically (usually at least monthly) to be assured that the water is free of pathogenic bacteria. City water is to flow into these tanks above the water level in order to attain the recommended air-gap separation. In addition, tanks are to be lined with a city code–specified lining, covered, and installed so that water will not be contaminated by pests, vermin, and unsanitary conditions. The potable water for the newer office buildings is supplied by demand-based booster pump systems connected to the city water main. Design calls for sizing the lead pump so the larger pumps are not required to cycle on and off. It is normally planned to provide 30-psig pressure at the top of a zone and not above 85-psig pressure at the bottom of a zone.[2] Cathodic protection may be installed on the potable water storage tanks to minimize corrosion. If the city potable water requires supplemental treatment in the building, it is necessary to refer and adhere to city codes.

Backflow prevention in drinking water systems[7,8]

While backflow prevention devices are necessary for any cross connection between a nonpotable water supply (such as cooling or closed system water) and potable water, installing these devices on fire protection systems has been practiced for almost a century.[7,9]

As the installation of in-line coffee makers may cause contamination of the potable water system, such installations are not allowed in building water systems.

Chilled water tanks

A nitrogen blanket is commonly installed on the top of the water level in expansion tanks to prevent additional oxygen from entering the system and increasing corrosion. Some large systems in which the piping is quite complex may experience high makeup water usage and as a result, even with proper treatment, may also experience appreciable corrosion. Since these systems may have a multitude of small valves and passageways, corrosion product accumulations (black iron oxide, magnetite) may make it necessary to install filter cartridges to filter out particles as small as 1 μm (see Chaps. 6 and 7).

The reason why high makeup water usage is sometimes encountered is often related to the high makeup water requirements during the period from 9:30 a.m. to 5 p.m. and minimal requirements after 5 p.m. Expansion tank capacities may not be adequate to take care of this fluctuating volume requirement of the system, and as a result, overflow and loss occur as contraction of the system volume occurs.

Fire protection systems

The older buildings included tanks holding about 50,000 gal in the fire protection system, but the newer systems include pumps connected directly to the city mains. Both kinds of systems were reported to be drained and filled once or twice a year and in general required minimum maintenance, consisting of attention to some "weeping" at flanges. Rather than draining the systems, fire departments may resort to testing the valves and pumps more frequently. As Chicago water is considered to be generally noncorrosive, it is not unexpected that minimum corrosion occurs in the fire protection systems. As mentioned previously, the application of Item 32 liquid sodium silicate at a dosage of 1/4 to 1/2 lb/1000 gal to a more corrosive water as the metal fire protection system is being filled will inhibit general corrosive tendencies. Of concern is the microbe-induced corrosion (MIC) that may develop in these stagnant anaerobic (oxygen-free) systems. Silicate treatment will be expected to raise the pH, reduce corrosion, and reduce bacterial growth conditions.

Despite variations in test results, the National Fire Protection Association (NFPA) Automatic Sprinkler Committee[9] has accepted plastic piping material in high-rise offices, hotels and apartments, hospitals, nursing homes, and places of assembly with many special installation conditions. Installation of plastic piping should eliminate the corrosion problem in new systems or in systems requiring replacement.[10-12]

One of the newer ideas for making better use of fire protection systems is to use their volume as a thermal energy system (TES). A TES helps meet heating or cooling demands and provides the benefit of lower electrical energy costs (at night), or it makes possible the installation of less costly equipment that would normally be needed to supply peak energy demands. It is questioned whether city or NFPA fire protection codes[9,13] would allow such designs to be installed.

References

1. E. G. Hansen, *Hydronic System Design and Operation,* McGraw-Hill, New York, 1985.

2. J. F. Mueller, *Plumbing Design and Installation Details,* McGraw-Hill, New York, 1987.
3. P. E. Beck, "Choosing the Right Plumbing System," *Consulting/Specifying Engineer,* June 1990, p. 52.
4. K. Johnson, "Commercial Building Energy Analysis Tools," *EPRI Journal,* April/May 1992, p. 30.
5. W. F. Mayer and R. Larsen, *Evaluation of White Rust and Cooling Tower Metallurgy,* Associated Laboratories, Palatine, Ill., 1992.
6. G. Duffy, "Humidity Control's Changing Frontier," *Engineered Systems,* May 1992, p. 19.
7. E. A. Towle and R. Ackroyd, "Protecting Drinking Water Systems from Backflow," *Consulting/Specifying Engineer,* November 1989, p. 64.
8. *Cross Connection Protection Devices,* 2d ed., American Society of Sanitary Engineers, Bay Village, Ohio, 1989.
9. *Inspection, Testing, and Maintenance of Sprinkler Systems,* NFPA 13A, and *Inspection, Testing, and Maintenance of Standpipe and Hose Systems,* NFPA 14A, National Fire Protection Association, Quincy, Mass., 1987, 1989.
10. R. G. Gewain, "Fire Sprinklers and Plastic Pipe?" *Southern Building,* May/June 1986.
11. J. B. Zicherman, "Is PVC Piping Firesafe?" *Fire Journal,* November/December 1990.
12. E. D. Yonkers, "Sprinklers Help Win the Fire Battle," *Consulting/Specifying Engineer,* April 1991, p. 32.
13. M. Meckler, "Integrated Fire Sprinkler and Thermal Storage Systems," *Heating/Piping/Air Conditioning,* May 1992, p. 67.

Appendix A

Water Treatment Specifications

Information that is of interest to all users of these specifications is as follows:

1. Supervisory service in the application or control of the various water treatment chemicals listed in these specifications is not included unless specified.
2. With respect to chemicals used for drinking water purposes, it shall be understood that Items 4, 5, 6, 7, 8, 14, 15, 18, 22, 23, 26, 31, 32, 44, 46, 47, 48, 49, 58, 65, 66, 73, 74, 75, 76, 107, 122, 127, L206, L209, and L238 shall contain no soluble mineral or organic contaminants in quantities capable of producing a deleterious or injurious effect on those consuming water that has been properly treated.
3. Chemicals purchased according to the following water treatment specifications shall be subject to inspection and testing before being applied for their specific use in plant operations. The supplier shall submit to the purchaser on delivery a signed certificate of composition for all the following chemicals. It shall be understood between the supplier and the purchaser that chemicals not meeting the specifications shall be subject to rejection.
4. All users of these chemicals shall adhere to the precautions in handling specified in the Material Safety Data Sheets submitted by the suppliers of these chemicals.

Item 1E *boiler water dispersant*. Partially desulfonated sodium lignosulfonate (combined CaO, MgO not more than 0.25%); as Maracell XE, manufactured by Reed Lignin, Inc., Rothschild, Wis., or equivalent.

Item 1F *boiler water dispersant and antifoam* shall consist of the following:

Item 1E	81–85%
Sodium sulfite, anhydrous (Item 10)	12–14%
Polyhydric alcohol type antifoam (Ucon 50-HB-5100)	3–4%

This blend shall be a uniform mixture, readily dispersible in water. Insoluble matter shall be less than 0.5%; pH of a 1% solution shall be above 7.4. It shall be specified to be supplied in 50-lb moistureproof bags or in fiber drums of 400-lb maximum size.

Item 4 *caustic soda* (NaOH), flake or beads, 76% Na_2O, in steel drums of 400 lb or 100 lb, as specified.

Item 5 *soda ash* (Na_2CO_3). It shall contain not less than 99% by weight sodium carbonate (Na_2CO_3), light, of which not less than 58% shall be in the form of sodium oxide (Na_2O). Insoluble matter shall be less than 0.05%. It shall contain no large lumps or chips or foreign matter that would make it unsuitable for storing in closed hoppers, transferring in conveyor equipment, or feeding in dry-feeding machines. It shall be supplied in 50- or 100-lb paper bags or as approved.

Item 6 *disodium phosphate, anhydrous* (Na_2HPO_4), white granules or powder, standard grade, in 100-lb moistureproof bags.

Item 7 *trisodium phosphate, monohydrate* ($Na_3PO_4 \cdot H_2O$), 38.5% P_2O_5, powdered, in 100-lb moistureproof bags.

Item 7A *trisodium phosphate, anhydrous* (Na_3PO_4), 43.3% P_2O_5, powdered, in 100-lb moistureproof bags.

Item 8 *sodium polyphosphate* (empirical formula of $Na_{12}P_{10}O_{31}$), 67% P_2O_5 (89.6% PO_4) or accepted equivalent, ground grade, in 100-lb moistureproof bags.

Item 9 *sodium nitrate* ($NaNO_3$), USP, granular or pellets, in 100-lb paper bags.

Item 10 *sodium sulfite, anhydrous* (Na_2SO_3), white crystals or powder, technical grade, containing 96–99% Na_2SO_3 and sufficient traces of iron and copper (>0.01%) to catalyze the oxygen-sulfite reaction in water. It shall not provide phenolic or objectionable odors in steam. Packed in 50-lb multiwalled paper bags.

Item 13 *cyclohexylamine* ($C_6H_{11}NH_2$), 60–65%, for inhibiting corrosion in condensate return lines. Softened water is to be used for dilution water. Packed in 5 gal (39+ lb net), 230 lb net, or 430 lb net, as specified.

Item 14 *hydrated lime* [$Ca(OH)_2$]. It shall contain not less than 90.0% $Ca(OH)_2$ (equivalent to 68% available CaO content) and shall disperse properly to provide satisfactory treatment or softening of public water supplies. It shall be white, dry, and free of lumps and any foreign material that may interfere with conveying or dry-feeding equipment. Packed in 50-lb multiwalled paper bags.

Item 15 *sodium aluminate* ($NaAlO_2$), containing a minimum of 65–68% soluble sodium aluminate, pulverized, 8% maximum insoluble, readily dispersible in water; 5% solution shall be stable for 24 h. Packed in 50-lb multiwalled paper bags.

Item 18 *tetrasodium pyrophosphate, anhydrous* ($Na_4P_2O_7$), white powder or granular, 53% P_2O_5, packed in moistureproof paper bags.

Item 20 *sulfuric acid,* 66° Baumé, H_2SO_4. In carboys of 5 gal (75 lb), 6 gal (90 lb), 6.5 gal (100 lb), or 13.5 gal (200 lb); in 55-gal drums (750 lb); or in bulk (tank truck) as specified.

Item 22 *quick lime* (CaO) shall contain not less than 90% available calcium oxide. It shall be of a quality known as "quick-slaking": it shall slake satisfactorily, readily disintegrating into a suspension of finely divided material without production of undissolved unslaked material in the slaker. It shall be substantially free of core, ash, and dirt; it shall pass through a 3/4-in screen, and not more than 5% shall pass through a No. 100 U.S. standard sieve; and it shall not contain material that may interfere with the operation of conveying or dry-feeding equipment. Packed in 50-lb multiwalled bags.

Item 23 *sodium tripolyphosphate, anhydrous* ($Na_5P_3O_{10}$), odorless, white powder or granular, 57% P_2O_5, packed in 100-lb moistureproof bags.

Item 25 *nitrogenous film-forming chemical,* for prevention of condensate return line corrosion. It shall contain not less than 15% active ingredient (for example, octadecylamine). The basic ingredient shall be a nitrogenous organic chemical of high molecular weight, which will vaporize through steam distillation and which is capable of developing a nonwettable film on the metal surface [for example, octadecylamine

($C_{18}H_{37}NH_2$)]. It shall have proven protective film-forming properties under condensate line conditions. Supplied in 350-lb (net) nonreturnable drums.

Item 26 *alum* [$Al_2(SO_4)_3$] shall contain not less than 17% available water-soluble alumina (Al_2O_3) and shall be basic to the extent of 0.3% by weight of water-soluble Al_2O_3 in excess of the amount theoretically required to combine with sulfur trioxide (SO_3) present, exclusive of that combined with foreign sulfates. Insoluble matter shall not exceed 0.5%. It shall be ground to such a size that not less than 90% shall pass through a Bureau of Standards No. 10 sieve, and 100% shall pass through a No. 4 sieve, and it shall be free of lumps and foreign material that may interfere with the operation of conveying and dry-feeding equipment. Packed in 50-lb multiwalled paper bags.

Item 29 *sodium metasilicate* (Na_2SiO_3), 51% Na_2O, anhydrous, granular, packed in 100-lb moistureproof paper bags.

Item 30 *morpholine* (C_4H_9ON), corrosion inhibitor grade, to be used for preventing corrosion in condensate lines. Morpholine content by weight shall be 40–42% by acid titration, maximum APHA color 15%. Packed in 450-lb drums.

Item 31 *monosodium phosphate* (NaH_2PO_4), anhydrous, standard grade, 58.5% P_2O_5, packed in 100-lb moistureproof bags.

Item 32 *liquid sodium silicate,* relatively low alkalinity, 41° Baumé, *approximately 28.8% SiO_2, 9.2% Na_2O*, alkali/silicate ratio 1:3.22, containing not more than 0.5% suspended matter. Packed in 60- or 600-lb drums.

Item 35 *zinc sulfate, monohydrate* ($ZnSO_4 \cdot H_2O$), white, free-flowing powder, soluble in water, packed in 100-lb. multiwalled paper bags.

Item 38 *sulfamic acid* (HSO_3NH_2), 99%, granular, free-flowing powder, for scale removal, packed in 100-or 400-lb. fiber drums.

Item 39 *corrosion inhibitor for sulfamic acid cleaning,* containing a mixture of several alicyclic amines of the abietyl amine type, ethylene oxide adducts of these amines, an ethylene oxide adduct of nonylphenol, approximately 5% dibutylthiourea, and approximately 30% isopropanol as a solvent; such as Inhibitor 98 manufactured by McGean Chemical Co., Cleveland, Ohio, or equivalent. At 1% concentration at least 90% inhibition efficiency shall be obtained on 1020 steel and 316 stainless steel at 140–160°F (60–71°C), and at least 50% inhibition efficiency on copper alloys. Recommendations call for the use of 1 lb inhibitor per 100 lb of Item 38 sulfamic acid. Packed in 5-gal (42.5-lb) nonreturnable containers.

Item 40 *hydrochloric acid* (*muriatic*) (HCl), technical, 18° Baumé, 27.9%, sp gr 1.14, for scale removal. Packed in carboys of approximately 6.5 gal (62 lb) or 13 gal (124 lb).

Item 41 *corrosion inhibitor for hydrochloric acid cleaning,* containing 55% of a 5-mol ethylene adduct of technical dehydroabietylamine to which is added 15% of the unsubstituted amine; all is dissolved in isopropyl alcohol; such as Hercules Powder Co. Polyrad 0515A or equivalent. In a 4-h test at 165°F (74°C) the coupon weight loss of 1010 mild steel exposed to 15% hydrochloric acid in presence of 0.1% inhibitor shall be reduced from a corrosion rate of 23.3 to 0.15 (in mils penetration per year). Recommendations call for the use of 1 lb of inhibitor to 100 lb hydrochloric acid (18° Baumé). Packed in 5-gal (42-lb) nonreturnable containers.

Item 44 *chlorine,* liquid, Cl_2, shall be 99.5% pure by volume as obtained from vaporized liquid chlorine and shall be suitable for water and sewage treatment. Valve outlet threads and face shall be in good condition so connections can be made without leaks. Two gaskets shall be furnished with each cylinder. In 100-lb or 150-lb cylinders as specified.

Item 46 *calcium hypochlorite,* $Ca(OCl)_2$, shall be a white or yellowish-white granular powder, free from lumps, dirt, or foreign material;, containing approximately 70% available chlorine by weight. Packed in 5-lb cans or 100-lb metal drums.

Item 46A *calcium hypochlorite tablets,* $Ca(OCl)_2$, active ingredient 65%. Each small (3/4-in-diameter) tablet weighs about 1 oz and when dissolved in 100 gal water is expected to provide a test of 10 mg/L available Cl_2; a larger (2 5/8-in-diameter) dissolving tablet weighs about 5 oz, such as Clarmor tablets or equivalent; purchaser shall specify size of tablet desired. Available in cases of four 10-lb containers and in 25- and 50-lb plastic pails.

Item 47 *sodium hypochlorite solution* (NaOCl) shall be a clear, light yellow liquid containing not less than 12.5% available chlorine by volume. Insoluble matter shall be less than 0.15% by weight. Total free alkali (such as NaOH) shall not exceed 1.5% by weight. Density is 12 lb/gal. Available in 1-gal or 5-gal containers or in 55-gal (600-lb) drums.

Item 48 *sodium zeolite,* polystyrene resin type, manufactured by copolymerizing styrene with divinylbenzene, followed by sulfonation. It shall have a water-softening capacity of 20,000 gr/ft^3 when regenerated with 6 lb of salt per cubic foot. A wet-screen analysis shall indicate 98% to pass through a 16-mesh screen and less than 1% to be retained on a 50-mesh screen. It shall be stable to 250°F (121°C), chemically stable throughout the entire pH range, and stable to oxidizing agents. Its physical form shall be hard, attrition-resistant, beadlike particles, shipped in moist, completely swollen condition. Packed in bags containing 1 ft^3.

Item 49 *powdered activated carbon* shall be a grade in extensive use for removal of taste and odor from municipal water supplies. It is a charred or carbonized residue of an organic material, prepared in such a way to make it possess a highly adsorptive capacity, particularly with respect to taste and odor-producing substances in water. The moisture shall be less than 8%. Fineness shall be such that not less than 99% will pass through a 100-mesh sieve, not less than 95% will pass through a 200-mesh sieve, and not less than 90% will pass through a 325-mesh sieve, as tested by the wet-screen method. Adsorption capacity shall be determined by the use of the standard phenol test or by measurement of the threshold odor reduction in the water to be treated. Packed in 35-lb multiwalled paper bags or approved equivalent.

Item 52 *copper sulfate* ($CuSO_4 \cdot 5H_2O$), at least 99% purity, blue, crystals sized about 1 in, packed in 100-lb multiwalled paper bags or approved equivalent.

Item 53 *inhibited (liquid) acid for removing scale* shall be composed of hydrochloric (or muriatic) acid (7–20° Baumé, 10–31.4% HCl) and a small percentage (approximately 1%) of an effective corrosion inhibitor. This formulation shall be effective in removal of calcium carbonate scale and shall provide at least 90% corrosion inhibition efficiency at 140–160°F (60–71°C) for 1020 steel and 50% inhibition efficiency for copper. Packed in 13-gal carboys or 55-gal corrosion-resistant drums.

Item 54 *inhibited (powdered) acid for removing scale.* It shall be composed of sulfamic acid ($HOSO_2NH_2$), 99% nonhydroscopic, and a small percentage of an effective corrosion inhibitor. This formulation shall be effective in removal of calcium carbonate scale and shall provide at 140–160°F (60–71°C) at least 90% corrosion inhibition efficiency for 1020 steel and 316 stainless steel and 50% inhibition efficiency for copper alloys. Packed in 100-lb corrosion-resistant drums or as specified.

Item 58 *slowly soluble sodium polyphosphate,* in glassy or lump form; contains calcium or magnesium to provide slower solubility in treating cooling waters to prevent scale formation. Approximate composition shall be either 68% P_2O_5, 16% CaO, and 20% Na_2O or 65% P_2O_5, 6.5% MgO, and 28% Na_2O. Dissolving rate shall be such that chemical may be added intermittently and that a uniform rate of soluble chemical will be provided. Solution rate shall be approximately 15% per month at 70–80°F (21–27°C). In 5-lb cans or 50-lb fiber drums.

Item 59 *sodium bisulfate* ($NaHSO_4$), globular, packed in 100-lb paper bags.

Item 60 *sodium pentachlorphenate* (C_6Cl_5ONa), technical, 90% minimum, in pellet form, 10–20 mesh or equivalent; packed in 100-lb drums.

Item 61 *sodium bichromate or sodium dichromate* ($Na_2Cr_2O_7 \cdot 2H_2O$), 99.8%, technical, granular. Packed in 100-lb paper bags.

Item 62 *hot water boiler treatment blend*
Borax ($Na_2B_4O_7 \cdot 5H_2O$)	5–12%
Sodium nitrite ($NaNO_2$), noncaking	65–70%
Sodium mercaptobenzotriazole, 50% aqueous solution (causticized $C_7H_5S_2$)	4–5%
Soda ash, light (Na_2CO_3)—a diluent for preventing caking	15–20%

The blend shall be a uniform mixture, readily soluble in water. Packed in 50-lb quantities in moistureproof bags or in 400-lb fiber drums of proven strength and chemical resistance as specified.

Item 65 *carbon dioxide gas,* commercial grade, guaranteed purity 99.5% CO_2; in 50-lb (438-ft^3) quantities in gas cylinders.

Item 66 *nitrogen gas,* water-pumped grade, 99.6% nitrogen, cylinder to contain 227 ft^3 and 16 lb of nitrogen.

Item 67 *hydrazine hydrate* ($N_2H_4 \cdot H_2O$), 25% hydrazine (N_2H_4) solution, sp gr 1.02 at 30°C; flash and fire point—none. To be used as oxygen scavenger in feedwater treatment. In nonreturnable 30-gal drums (250 lb net).

Item 68 *sodium salt of ethylenediamine tetracetic acid* (EDTA), containing approximately 38% tetrasodium ethylenediamine tetraacetate (such as Vertan 600, manufactured by Dow Industrial Service, Midland, Michigan, or equivalent). Packed in 600-lb drums.

Item 70 *corrosion inhibitor blend,* consisting of pH-adjusted sodium bichromate for use in closed hot and cooling water systems and dehumidifying systems, where chromate is allowed by the U.S. Environmental Protection Agency (EPA). It shall contain the following:
Sodium bichromate ($Na_2Cr_2O_7$) (Item 61)   60–62%
Soda ash (Na_2CO_3) (Item 5)   38–40%
The blend shall be a uniform mixture that readily dissolves in water. Insoluble matter shall be less than 0.1%. Packed in 50-lb (net) moistureproof bags or 400-lb (net) fiber drums of adequate structural strength and chemical resistance.

Item 72 *sodium nitrite* ($NaNO_2$), USP granular, applied for corrosion inhibition, 99.5% minimum purity; packed in 100-lb multiwalled paper bags.

Item 73 *ferric sulfate* [$Fe_2(SO_4)_3$], water-soluble ferric ion shall be not less than 18.0%; total insolubles shall not exceed 6.0%. Typical screen analysis indicates 100% should pass through 3-mesh; 95% through 4-mesh. For use as a coagulant in water, sewage, and industrial waste treatment. Packed in 100-lb polyethylene-lined bags.

Item 74 *potassium permanganate* ($KMnO_4$), 97% minimum, technical and free-flowing, crystalline. Typical screen analysis is for more than 20% by weight to be retained on a 40-mesh screen and no more than 20% to pass through a 200-mesh screen. For taste and odor control of potable waters and for algae control of cooling towers. Packed in 110-lb (net) steel drums.

Item 75 *slowly soluble sodium zinc polyphosphate.* In glass platelet form 1/16–1/4 in thick, with pieces not over 2 in by 3 in. Approximate composition shall be 60% P_2O_5, 10% ZnO, and 30% Na_2O. Complete dissolution shall be obtained in tap water within 7–15 h at 70°F (21°C). For scale and corrosion prevention in the smaller evaporative cooling and distribution systems. Packed in 100-lb (net) multi-ply, water-resistant (burlap-reinforced) bags.

Item 76 *diatomite,* diatomaceous silica, white powder, flux calcined, filter aid suitable for removal of fine suspended solids from swimming pool water during filtration; such as Celite 545 or equivalent. Packed in 50-lb multiwalled paper bags.

Item 84 *pentasodium aminotrimethylphosphonate* (Na_5ATMP), 39–41%, pale yellow solution, pH 10.0–11.0 in 1% solution, sp gr 1.42 (20°C/15°C). This chemical shall prevent the precipitation of hardness when used at threshold dosages (2–20 ppm) in cooling water; such as Dequest 2006, manufactured by the Monsanto Co., St. Louis, Missouri, or equivalent. Packed in 5-gal (55-lb) or 15-gal (170-lb) drums, polyethylene-lined, with steel or fiber 3 outer pack, or in 55-gal (600-lb) drums of the same type.

Item 93 *nitriloacetic acid trisodium salt, monohydrate,* minimum 99% active, $NTA.H_2O$, white granular powder, packed in 50-lb polyethylene-lined, multi-ply paper bags.

Item 95 *diphosphonic acid formula* (HEDP) (1-hydroxethylidene-1, 1-diphosphonic acid), for prevention of scale in cooling water, active ingredient 58–62%, sp gr 1.45 at 20°C/15°C, pH in solution <2.0; such as Dequest 2010 or equivalent. Packed in 15-gal (170-lb net) polyethylene-lined, nonreturnable steel drums or 55-gal (600-lb) polyethylene-lined, nonreturnable drums.

Item 102 *methylene-bis-thiocyanate biocide* (Y), for control of microorganisms in cooling water systems. This biocide degrades increasingly at pHs above 8–9 and accordingly shouldn't be used in cooling waters with a pH this high. It shall contain 10% methylene-bis-cyanate in water solution. Packed in 55-gal (450-lb) drums or in smaller sizes as specified.

Item 103 *carboxylic cation exchange resin,* weakly acidic, theoretical capacity of 4.25 mequiv/mL (93-kg capacity as $CaCO_3$ per cubic foot), 43–50% moisture; effective size 0.38–0.47 mm, uniformity coefficient of 1.75 max., pK of about 5.3 with respect to Na in a 1.0-*M* solution; such as Rohm & Haas Amberlite IRC84 or equivalent.

Item 104 *aminotri (methylene-phosphonic acid)*, $N(CH_2PO_3H_2)_3$, for prevention of scale in cooling water; active ingredient 49–51%, sp gr 1.32–1.34, pH of 1% solution is 2.0 at 25°C; such as Dequest 2000 or equivalent. Packed in 15-gal (166-lb) polyethylene-lined containers with steel or fiber outer pack.

Item 107 *tetrapotassium pyrophosphate, anhydrous* ($K_4P_2O_7$), pellet form, 42.7% P_2O_5, solubility 188 g/100 g H_2O at 25°C. Packed in 50-lb moistureproof bags.

Item 108 *biocidal agent blend* (Z) for controlling slime-forming microorganisms in cooling water systems. This liquid formulation [pH >10.5, sp gr 0.98 (65°F, 18°C)] applied at the rate of 0.1–0.75 lb/1000 gal system volume shall be effective in controlling slime and shall contain:

N-alkyl (C_{12}—5%, C_{14}—60%, C_{16}—30%, C_{18}—5%)
Dimethyl benzyl ammonium chloride	23–25%
Bis (tributyltin) oxide	4–6%
Inert ingredients, such as solubilizing and dispersing agents	70–72%

To be supplied in 5-gal (40-lb) nonreturnable steel pails with pour spouts.

Item 112 *causticized mercaptobenzotriazole* ($C_7H_4NS_2Na$), corrosion inhibitor for copper, 50% aqueous caustic solution; such as Nacap or equivalent; packed in 5-gal (40-lb) steel containers.

Item 113 *benzotriazole* ($C_6M_4N_2NH$), powder, corrosion inhibitor for copper; such as Cobratec 99 or equivalent; packed in 200-lb fiber drums.

Item 116 *diethylethanolamine* [$(C_2H_5)_2N.CH_2CH_2.OH$], corrosion inhibitor grade for inhibiting corrosion in condensate lines, 99.5% minimum, maximum APHA color 15, sp gr @ 20°/20°C is 0.88–0.89. Packed in 55-gal (400-lb) drums.

Item 117 *catalyzed hydrazine hydrate* ($N_2H_4.H_2O$), 35% hydrazine, sp gr 1.02 at 30°C, flash and fire point—none; containing an organic catalyst that increases the reaction rate to at least 10 times that of uncatalyzed hydrazine when used as an oxygen scavenger or corrosion inhibitor in boiler feedwater treatment or in steam condensate line treatment when fed to the steam drum; such as Drew Amerzine 35 or equivalent. Packed in 53-gal (441-lb net) drums.

Item 119 *softened water boiler treatment,* powdered, blend L
Organic blend (Item 1F)	20–25%
Disodium phosphate (Item 6)	15–20%
Sodium sulfite (Item 10)	57–62%

This blend shall be a uniform mixture that readily dissolves in water and does not contain more than 0.5% insoluble matter. Packed in 50-lb quantities in moistureproof bags or 400-lb fiber drums as specified.

Item 122 *sodium acid pyrophosphate* ($Na_2H_2P_2O_7$), 63.5% P_2O_5, white powder; packed in 100-lb moistureproof bags.

Item 123 *sodium sulfite, pure,* food grade, for reacting with dissolved oxygen to reduce corrosion in water systems; packed in 100-lb moistureproof bags.

Item 124 *sodium sulfite, catalyzed,* containing sufficient cobalt or some other equally effective catalyst to increase the reaction of sulfite and dissolved oxygen to over 20 times faster; packed in 100-lb moistureproof bags.

Item 125 *silicone-type antifoam,* food grade, antifoam A content 10.5%, silicone content 4.0%, cream appearance, for defoaming fountains or cooling basins at 50 ppm dosage; such as Dow Corning FG10 or equivalent. Packed in 5-gal (40-lb net) containers.

Item 127 *liquid caustic soda* (NaOH), 48.5% minimum [should be stored above 50°F (10°C) to prevent crystallization]. Packed in 55-gal drums containing 700 lb net.

Item 130A *algaecide A*
  Poly(oxyethylene(dimethyliminio)ethylene
    (dimethyliminio)ethylene dichloride)   60%
  Inert ingredients   40%
A 9.5 lb/gal water-soluble liquid, such as Buckman Lab WSCP or equivalent. Packed in 5-gal (46-lb), 15-gal (138-lb), or 30-gal (276-lb) containers.

Item 131A *algaecide B*
  Disodium cyanodithioimidocarbamate   14.7%
  Potassium *N*-methyldithiocarbamate   20.3%
  Inert ingredients   65.0%
A 10.2 lb/gal water-soluble liquid, such as Buckman Lab NABE-M microbiocide concentrate. Packed in 5-gal (48-lb), 15-gal (134-lb), or 30-gal (189-lb) containers.

Item 147A *sodium tolyltriazole* ($C_7H_6N_3Na$), 50% water solution, sp gr 1.19 at 25°C, clear, red-brown, essentially colorless, pH 13.5; such as

Cobratec TT-50S, from PMC Specialties Group, Cleveland, Ohio. Packed in 55-gal standard drums (500 lb net) or in 40-lb drums.

Item 148 *low-pressure boiler water treatment blend*
Soda ash (Na_2CO_3)	28–32%
Disodium phosphate, anhydrous (Na_2HPO_4)	28–32%
Sodium sulfite, anhydrous (Na_2SO_3)	22–26%
Partially desulfonated sodium lignosulfonate (as Item 1E)	9–13%
Sodium polyphosphate, Item 8	4–8%

This blend shall be a uniform mixture, readily dispersible in water. In 50-lb moistureproof bags or in fiber drums containing a maximum of 400 lb.

Item 149 *low-pressure boiler liquid water treatment*
Sodium sulfite (Na_2SO_3)	4–6%
Sodium tripolyphosphate ($Na_5P_3O_{10}$)	1.5–2.5%
Disodium phosphate, anhydrous (Na_2HPO_4)	7–9%
Potassium hydroxide (KOH)	4–6%
Softened or deionized water	76–80%

This liquid treatment shall contain less than 0.1% insoluble matter. Packed in 30-gal (est. 290-lb) rotationally molded drums with fiber Overpak.

Item L200 *sodium polyacrylate,* powder, neutralized, molecular weight 2800, pH 5.5–6.5, 10–15% moisture (such as Goodrich Chemicals K759 or equivalent); packed in 55-gal drums (500 lb net) or in smaller quantities at increased cost.

Item L206 *calcium hypochlorite* in tablet form [$Ca(OCl)_2$]. These tablets shall contain not less than 65% available chlorine by weight and shall dissolve in 2–4 hr at 85°F (30°C). One ounce per 5000 gal provides a dosage of approximately 1 ppm of available chlorine.

Item L208 *liquid chromate solution* for closed systems
Sodium dichromate (Item 61)	24–26%
Caustic soda (Item 4)	5–7%
Softened water	65–70%

This liquid shall contain less than 0.1% insoluble matter. Packed in 55-gal drums.

Item L209 *liquid sodium metaphosphate,* $(NaPO_3)_6$, containing a trace amount of sodium hypochlorite to inhibit bacterial growth, pH 6.6–7.0. This item is easily soluble in water. It is to be supplied in corrosionproof drums.

L211 *sodium acid sulfite,* $NaHSO_3$, used as an oxygen scavenger when an acid product is desired. Supplier is to advise composition of product being supplied. Supplied in corrosionproof drums.

L212 *sodium molybdate,* $(Na_2MoO_4 \cdot 2H_2O)$, crystalline, in 200-lb net weight fiber drums.

L213 *acrylate copolymer-1.* Effective as a stabilizer of phosphate-based cooling tower treatment programs (such as Rohm & Haas Acumer 2000 or equivalent), active solids 39.5%, mol wt 3300–4500, pH 4.2, density 10.1 lb/gal, in 525-lb net weight corrosionproof drums.

L214 *acrylate copolymer-2.* Effective as a calcium phosphate scale inhibitor and dispersant additive in cooling water treatment programs (such as Good-rite K-796), 45% total solids, pH 4.5, sp gr 1.17 at 25°C, supplied in 500-lb net weight corrosionproof drums.

L215C *phosphate-copper inhibitor* for cooling water treatment, composed of the following:

Item 7A trisodium phosphate, anhydrous	48–54%
Item 107 tetrapotassium pyrophosphate, anhydrous	40–45%
Item L221 sodium tolyltriazole (Cobratec TT-85)	11–12%

This blend shall be a uniform dry mixture that readily dissolves in water, shall not contain >0.5% insoluble matter in water, and shall provide a clear, homogeneous solution when added first and mixed well in a full vat of water followed by the addition of L220 in preparing a combined 5% solution for application by a typical chemical feeding system. Packed in 100-lb or 500-lb moistureproof containers.

L217A *liquid acrylate-phosphonate-phosphate-inhibitor* treatment for cooling towers; 0.5 lb/1000 gal makeup shall be applied to inhibit scale and corrosion. Composition is as follows:

Softened or deionized water	61–65%
Item 95 diphosphonic acid	7–8%
Item 218A acrylate terpolymer	12–13%
Item 7A trisodium phosphate, anhydrous	6–7%
Item 107 potassium pyrophosphate ($K_4P_2O_7$)	6–7%
Potassium hydroxide (KOH), sufficient to raise pH to 11.0	2.5–3%
Item 147A sodium tolyltriazole	1.5–2%

Add these ingredients in the order listed above. This liquid treatment solution shall be clear and homogeneous and packed in 55-gal corrosionproof drums.

L218A *terpolymer for dispersing boiler water sludge.* This acrylic terpolymer (such as Rohm & Haas Acumer 3100 or equivalent), which effectively disperses dried iron oxide and hydrated iron oxide, has the following physical properties: total solids 43–44%, sp gr 1.10, pH 2.5–3.0. In neutralized form, it is to be maintained at 5–15 mg/L as a dispersant in boiler water and is packed in 55-gal corrosionproof drums.

L219 *terpolymer-phosphate boiler water treatment.* This neutralized boiler water treatment blend provides a terpolymer for iron oxide scale inhibition and phosphate for scale inhibition and shall be composed of the following:

Softened or deionized water	69–71%
Ucon 50-HB-5100, Union Carbide boiler antifoam	1.5–3%
Item L218A terpolymer	12–16%
Item 4 caustic soda	2–4%
Item 8 sodium polyphosphate	9–11%

The solution shall be clear and homogeneous with a pH of about 12.0. It shall be packed in 55-gal corrosionproof drums or smaller-sized containers as requested.

L220 *liquid phosphonate-terpolymer mix* for cooling water treatment shall be composed of the following:

Item 95 diphosphonic acid	38–43%
Item L218A polymer (terpolymer)	34–36%
Softened or deionized water	24–26%

Insoluble matter shall be <0.5%. This liquid blend shall provide a clear, homogeneous water solution when a combined 5% solution of L220 and L215C (2:1 ratio) is prepared in a chemical vat for application by a typical chemical feeding system. Packed in 55-gal corrosionproof drums.

Item L220A *liquid phosphonate-terpolymer mix requiring lower phosphonate cooling water treatment* shall be composed of the following:

Item 95 diphosphonic acid	24–26%
Item L218A polymer (terpolymer)	53–57%
Deionized water	19–21%

Insoluble matter shall be <0.5%. This liquid blend shall provide a clear, homogeneous water solution when a combined 5% solution of L215C and L220A (1:2 ratio) is prepared in a chemical vat for application by a typical chemical feeding system. Packed in 55-gal corrosionproof drums.

Item L221 *tolyltriazole sodium salt* ($C_7H_6N_3Na$), 92–95%, containing less than 0.5% NaOH; off-white to light tan powder, mol wt 155.16 (such as Cobratec TT-85); packed in 150-lb fiber drums.

Item L222 *phosphonate (PBTC) (Bayhibit AM)*, active agent 2-phosphonobutane-1,2,4-tricarboxylic acid, empirical formula $C_7H_{11}O_9P$, mol wt 270, P content 11.5%, clear, colorless to light yellow liquid, density 1.27–1.30 g/cm^3, pH of 1% water solution = 1.5–1.8. For control of scale and corrosion in cooling water systems. Packed in 550-lb plastic drums.

Item L223 *phosphonate (HPA) (Belcor 575)*, hydroxy phosphonic acid, $PO(OH)_3CHCOOH$, mol wt 156, approximately 50% water solution, pH of 1% solution <1.5, sp gr 1.32–1.42. For control of scale and corrosion in cooling water systems. Packed in 55-gal corrosionproof drums.

Item L224 *deposit control agent (Belcor 283)*, calcium carbonate threshold inhibitor, general dispersant and stabilizer for L223 (Belcor 575) and zinc, mol wt 1200, pH of 1% solution <2.0; packed in 55-gal corrosionproof drums.

Item L225 *all-organic cooling water treatment,* composed of the following:

Water, deionized or softened	55–60%
Item 4 caustic soda	10–12%
Item 147A sodium tolyltriazole	2–3%
Item L218A terpolymer	17–20%
Item 95 diphosphonic acid	3.5–5%
Item L222 phosphonate (PBTC)	4–6%

This mixture shall be a clear solution containing <0.5% insoluble matter, shall be prepared by mixing in the above order, and shall have a pH of 12.5+. Sufficient amount shall be applied to provide 60–100 mg/L in the circulating water in order to control scale and corrosion in the cooling water system. It shall be packed in 55-gal corrosionproof drums.

Item L226 *molybdate-organic mixed cooling water treatment,* composed of the following:

Water, deionized or softened	45–50%
Item 4 caustic soda	9–11%
Item 147A sodium tolyltriazole	4–5%
Item L218A terpolymer, Acumer 3100	9–11%
Item 95 diphosphonic acid	4–6%
Item L222 phosphonate (PBTC)	3.5–4.5%
Item L212 sodium molybdate	19–21%

This mixture shall provide a clear solution containing <0.5% insoluble matter, shall be prepared by mixing in the above order, and shall have a pH of >12.5. Sufficient amount shall be applied to provide at least 50 mg/L in the makeup water, with pH above 8.2 (P alkalinity 20–60 mg/L, M alkalinity 200–300 mg/L) and molybdate (MoO_4) test of 10–15 mg/L in order to control scale and corrosion in the cooling tower water. It shall be packed in 55-gal corrosionproof drums.

Item L227 *molybdate-zinc liquid cooling tower treatment* is composed of the following:

Softened water	60–65%
L218A terpolymer, Acumer 3100	7–8%
Item 95 diphosphonic acid (HEDP)	3.5–4.5%
Item L222 phosphonate (PBTC)	3–4%
Item 212 sodium molybdate	7–8%
Item 35 zinc sulfate, monohydrate	1.5–2.5%

When these ingredients are completely dissolved, add:

Item 4 caustic soda	7–8%
Item 147A sodium tolyltriazole	3–4%

This mixture shall provide a clear solution containing <0.5% insoluble matter and shall have a pH >12.5. Sufficient amount shall be applied to provide >60 mg/L in the makeup water and to provide a pH of >8.2 (for example, P alkalinity of 20–60 mg/L, M alkalinity of 200–300 mg/L) and a molybdate (MoO_4) test of 7–12 mg/L in order to control scale and corrosion in the circulating water. It shall be packed in 55-gal corrosionproof drums.

Item L229 *cooling tower biocide (M)*, for control of algae and bacteria in recirculating cooling water systems. A broad-spectrum bromine-chlorine-organic complex biocide effective to pH of 9.0 containing:

1-brom-3-chloro-5,5-dimethylhydantoin	96%
Inert ingredient	4%

Recommended dosage is 0.1 to 0.3 lb per 1000 gal of water in the system to maintain 1.0 mg/L of bromine residual for at least 4 h (such as Western Chemical Company's Bromicide TM, EPA Registration Number 5785-57-7547; HOH Chemicals Co. Bromicide-T; or equivalent). Supplied in stick, tablet, or briquette form and packed in approximately 50-lb net weight fiber cartons.

Item L230 *cooling water organic blend to be used in preparing molybdate and phosphate formulations* (note, it contains no zinc). This blend, composed of phosphonates and a polymer, acts as a scale inhibitor and dispersant in providing scale-free surfaces. It shall be composed of the following:

Softened or deionized water	24–26%
Item L218A polymer (terpolymer)	34–36%
Item 95 diphosphonic acid	19–21%
Item L222 phosphonate (PBTC)	19–21%

Insoluble matter shall be <0.5%. This liquid blend shall provide a clear, homogeneous solution when mixed with Item L243 and Item 147A in a 1:0.4:0.1 ratio in the preparation of a 5% total solution for a typical chemical feeding system. Total treatment (L230, L243, and 147A) is based on feeding a total of 0.25 lb/1000 gal of makeup water (30 mg/L) and the cooling tower operating at 4 cycles of concentration. Packed in 55-gal corrosionproof drums.

Item L231 *sodium tolyltriazole–caustic soda blend.* This mixture shall provide a clear solution containing <0.5% insoluble matter. It shall be composed of the following:

Softened or deionized water	24–26%
Item 127 liquid caustic soda	50–52%
Item 147A sodium tolyltriazole	22–24%

When mixed with L230, L233B, or L234A, this shall provide a soluble chemical vat solution for attaining desired corrosion-inhibiting properties of the triazole. Packed in 55-gal or smaller-sized steel drums.

Item L232 *sodium bromide liquid,* 40% aqueous solution, clear liquid, density 1.34–1.51 g/mL, neutral, approximately 318 mL/lb. When used in a ratio of 0.02 gal of L232 to 0.03 gal of Item 47 sodium hypochlorite, provides effective microbiological control of cooling water. Available in 550-lb nonreturnable drums or in bulk. (Such as Buckman Labs. BULAB 6040 or equivalent.)

Item L233B *cooling water organic blend to be used in preparing molybdate, phosphonate, and zinc formulations.* This liquid blend, when used in combination with L231, shall provide scale inhibition and corrosion inhibition of steel and copper in cooling water systems. It shall be composed of the following:

Softened or deionized water	19–20%
Item 218A polymer (terpolymer)	21–24%
Item 95 diphosphonic acid	12–14%
Item L222 phosphonate (PBTC)	12–14%
Item L243 sodium molybdate, liquid	28–32%
Item 35 zinc sulfate, monohydrate	3–4%

Insoluble matter shall be <0.5%. This liquid blend shall provide a clear, homogeneous solution when mixed with Item L231 in a 1:0.2 ratio in the preparation of a 5% total solution for a typical chemical feeding system. It is estimated that 0.28 lb/1000 gal of makeup water (34 mg/L) is the dosage required to provide proper treatment of a tower operating at 4 cycles of concentration. Packed in 55-gal corrosionproof drums.

Item L234A *cooling water organic blend to be used singly or in combination with copper inhibitor and/or molybdate.* This blend, composed of phosphonates and zinc, acts as a scale and corrosion inhibitor and dispersant in providing scale-free and corrosion-free surfaces. It shall be composed of the following:

Softened or deionized water	26–27%
Item L218A polymer (terpolymer)	31–33%
Item 95 diphosphonic acid	17–19%
Item L222 phosphonate (PBTC)	17–19%
Item 35 zinc sulfate, monohydrate, or L247 zinc chloride to provide 2.0% zinc	4–6%

Insoluble matter shall be <0.5%. This liquid blend shall provide a clear, homogeneous solution when mixed with Item L243 and Item 147A in a 1.0:0.6:0.1 ratio in the preparation of a 5% total solution for a typical chemical feeding system. Application of 0.27 lb/1000 gal (32 mg/L) of the total solution for the makeup water should provide proper treatment at 4 cycles of concentration in the cooling tower. Packed in 55-gal corrosionproof drums.

Item L236 *liquid alkaline-terpolymer-antifoam boiler water treatment* shall be composed of the following:

Item 218A terpolymer	59–61%
Item 4 caustic soda	10–12%
Ucon 50-HB-5100, Union Carbide boiler antifoam	1–3%
Softened or deionized water	20–30%

The solution shall be clear and homogeneous with a pH of about 12.0. It shall be packed in 55-gal corrosionproof drums or smaller-sized containers, as specified.

Item L237 *glutaraldehyde* ($C_5H_8O_2$), mol wt 100.11; liquid nonoxidizing microbiocide, used at 10–75 mg/L intermittent feed to cooling towers for control of bacteria, fungi, and algae; effective over a broad pH range.

Item L238 *zinc orthophosphate liquid* formulation for inhibiting corrosion of lead, steel, and copper-bearing metals in distribution water systems; $Zn:PO_4$ ratio 1:1; application of 5–10 gal per 1,000,000 gal shall provide 0.5–1.0 mg/L zinc in treated water; properties: pH <1.0, density 10.6 lb/gal, 583 lb/55-gal drum. (Available as V932 from Technical Products Corp., Portsmouth, VA 23704, or equivalent.)

Item L239 *terpolymer for dispersing boiler water sludge*. This is a clear, thermally stable, liquid copolymer, such as Rohm & Haas Acumer 2100 Copolymer or equivalent. FDA-approved; can effectively disperse iron oxide; properties: mol wt 11000, active solids 31.0%, pH 4.8. In neutralized form, it is to be maintained at 5–15 mg/L as a dispersant in boiler water. Packed in 55-gal corrosionproof drums.

Item L240 *liquid alkaline-terpolymer-antifoam boiler water treatment* shall be composed of the following:

Item L239 terpolymer, FDA-approved	60–62%
Item 4 caustic soda	9–11%
Union Carbide 50-HB-5100 boiler antifoam	1.5–3%
Softened or deionized water	28–30%

The solution shall be clear and homogeneous with a pH of about 12.0. It shall be packed in 55-gal drums or smaller-sized containers, as specified.

Item L240A *liquid alkaline-terpolymer-antifoam boiler water treatment* shall be of the same composition as L240, but L218A copolymer shall be substituted for L239 copolymer.

Item L241 *nonionic nonfoaming surfactant* for use with 2 lb/1000 gal Item 31 sodium acid phosphate for surface action in passivating new galvanized-steel cooling towers. This surfactant applied at 10 mg/L dosage with Item 31 shall have the desired properties of surface action. In combination with caustic soda or sodium hypochlorite, it shall also be useful for more effective chemical cleaning. (Such as Rohm & Haas Surfactant Triton CF-54 or equivalent.)

Item L243 *sodium molybdate, 35% solution,* sp gr 1.40 min.; 100 lb contains 41.13 lb $Na_2MoO_4 \cdot H_2O$ or 28.35 lb molybdate (MoO_4) or 16.45 lb molybdenum (Mo); packed in quantities of 600 lb net in 55-gal steel drums.

Item L244 *penetrant and dispersant for microbial and organic deposits,* 18-carbon unsaturated carboxylic acid, such as N,N-dimethylamides, density 0.90 g/mL, approximately 505 mL/lb. Effective at 50–100 mg/L in cleaning, or at 10–20 mg/L normal usage with biocide. Available in 400-lb nonreturnable drums. (Such as Buckman Labs. BULAB 8007 or equivalent.)

Item L245 *sodium tetraborate* ($Na_2B_4O_7 \cdot 5H_2O$), 99.5%, white, granular; packed in 100-lb paper bags.

Item L246 *molybdate closed system treatment* composed of sodium molybdate, copper inhibitor, borax, and soda ash; applied at a dosage of 14 lb/1000 gal will inhibit corrosion of copper and steel in closed water systems. Composition shall be as follows:

Zeolite-softened water	49–51%
Item L243 liquid sodium molybdate	29–31%
Item L245 sodium tetraborate	14–16%
Item 147A sodium tolyltriazole	0.7–0.9%
Item 5 soda ash (sufficient to provide pH of 9.0 in closed water system)	2.5–5.0%

L247 *zinc chloride,* white, odorless, deliquescent, at least 95%.

L248 *phosphonate-polymer-zinc* formulation for inhibiting scale and corrosion when applied in a 0.5–1 lb/1000 gal dosage in cooling water systems. The composition shall be as follows:

Softened water	32–34%
L222 phosphonate, such as Bayhibit AM	21–23%
L218A terpolymer, such as Rohm & Haas Acumer 3100	43–45%
L247 zinc chloride	0.6–0.8%

This formulation shall be free of insoluble matter and shall readily disperse in dilution water. It shall be packed in 55-gal corrosionproof drums.

L249 *catalyzed oxygen scavenger* [*catalyzed diethylhydroxylamine (DEHA)*], 25% active, (such as patented product available as Neutrox 53 from Dearborn Grace, Lake Zurich, Ill.); applied at 50-ppb feedwater dosages in electrical utility high-pressure plants, 500–2000 ppb in industrial plants; used in conjunction with neutralizing amines; packed in 30-gal drums weighing 248 lb or 55-gal drums weighing 455 lb.

L250 *anion exchange resin,* strongly basic, Type II, quaternary ammonium resin, for use in dealkalizer system for alkalinity reduction, 44 lb/ft^3, 38–43% moisture; shipped in chloride form as uniform, attrition-resistant spherical particles, 20–50 mesh, less than 3% fines; such as Rohm & Haas Ira-400 or Ionac ASB-2.

Appendix B

Specifications for Water-Testing Reagents and Equipment

Information of interest to all users of these specifications is as follows:

1. Supervisory service in the application or control of the various water-testing reagents and equipment listed in these specifications is not included unless specified.
2. Items purchased according to the following specifications for water-testing reagents and equipment shall be subject to inspection and testing before being applied for their specific use in plant operations. The supplier shall submit to the purchaser on delivery a signed certificate of composition for all the following chemicals. It shall be understood between the supplier and the purchaser that chemicals and equipment not meeting the specifications shall be subject to rejection.
3. All users of these chemicals shall adhere to the precautions in handling specified in the Material Safety Data Sheets submitted by the suppliers of these chemicals.

Item No.   **Specifications**

Standard hardness titrating solution, containing sodium salt of ethylenediamine tetraacetic acid (1 mL = 1 mg $CaCO_3$) for determination of hardness. Solution shall retain its standard concentration for period of 1 year.

5002   In quart containers.
5003   In 1-gal containers.

Hardness buffer solution, to be used with standard hardness titrating solution (Item 5002 or 5003) and hardness indicator powder (Item 5014) for the determination of hardness content. This solution is to contain a buffer for maintenance of pH of 10 during the titration. Solution shall be of such strength that 1 mL added to a 50-mL water sample provides satisfactory analytical results.

5011   In pint containers.

Hardness indicator powder, to be used with standard hardness titrating solution (Item 5002 or 5003) and hardness buffer solution (Item 5011) for the determination of hardess content. This powder is to contain Eriochrome Black T (F241) of such concentration that 0.15 g added to a 50-mL water sample provides satisfactory analytical results.

5014   In 4-oz containers.

Standard sulfuric acid solution, $N/50$ $(0.02\ N)$, for titration of alkalinity.

5021   In quart containers.
5022   In 1-gal containers.

Methyl orange indicator solution, approximately 0.05–0.1%, for determination of total alkalinity.

5029   In pint containers.

Phenolphthalein indicator solution, approximately 0.5–1.0%, for determination of phenolphthalein alkalinity.

5033   In pint containers.

Standard silver nitrate solution (1 mL = 1 mg Cl^-), $0.0282\ N$, for titration of chloride.

5041   In quart containers.
5042   In 1-gal containers.

Hydrogen peroxide solution, 3%, for eliminating the interference of excess sodium sulfite in the determination of chloride in a boiler water sample.

5043   In 8-oz containers.

Potassium chromate indicator solution, approximately 5% for indicator, for determination of chloride.

5047   In pint containers.

5048   In quart containers.

Gallic acid, dry, crystals, pure, for neutralizing phenolphthalein alkalinity in the determination of dissolved solids by electrical conductivity.

5057   In 1-lb containers.

Standard potassium iodide-iodate solution (1 mL = 1.0 mg SO_3^{2-}), 0.025 N, for titration of sulfite content.

5067   In quart containers.
5068   In 1-gal containers.

Dual-purpose sulfite indicator powder, containing pregelatinized water-soluble starch and sulfamic acid. Less than 1 g of this reagent shall be sufficient to provide accurate analytical results in the determination of the sulfite content of boiler water using potassium iodide-iodate as the titrant. Minimum shelf life of this reagent shall be 8 to 12 months.

5072   In 1-lb containers.

Phosphate indicator powder, composed of finely ground stannous chloride powder, finely ground salt to provide bulk, and an anticaking agent. A quantity of 0.2 g is applied when the 0–100 ppm (PO_4) comparator (Item 5159) is used, and 0.1 g is applied when the 0–25 ppm (PO_4) comparator (Item 5165) is used; then, when the powder is properly dissolved in the water sample and Item 5084 is applied in the proper quantity, the correct phosphate results shall be obtained. The 2-oz quantity shall be adequate for determining phosphate content in 280 boiler water samples.

5075   In 2-oz containers.

Molybdate reagent solution, required along with Item 5075 for determining phosphate content when Taylor Chemicals, Inc., comparator (Items 5159 and 5165) is used.

5083   In quart containers.
5084   In 1-gal containers.

Sodium hydroxide solution, 1.0 N, for determining calcium with calcium indicator powder (Item 5099) and standard hardness titrating solution (Item 5002 or 5003).

5095   In pint containers.

Calcium indicator powder, composed of ammonium purpurate (murexide) and inert ingredients, for the determination of calcium content. When 0.15 g of this reagent and 2 mL of sodium hydroxide solution (Item 5095) are applied to a 50-mL water sample, titration with standard hardness titrating solution (Item 5002 or 5003) shall yield satisfactory calcium results.

5099   In 4-oz containers.

Meta-cresol purple pH indicator solution, for pH determination in the range of 7.6–9.2 (solution adjusted to 8.4) in Taylor Chemicals, Inc., comparator with Item 5167 slide.

5105   In 4-oz containers.

Phenol red pH indicator solution, for pH determination in the range of 6.8–8.4 (solution adjusted to 7.6) in Taylor Chemicals, Inc., comparator with Item 5173 slide.

5106   In 4-oz containers.

Bromthymol blue pH indicator solution, for pH determination in the range of 6.0–7.6 (solution adjusted to 6.8) in Taylor Chemicals, Inc., comparator with Item 5172 slide.

5107   In 4-oz containers.

Thymol red pH indicator solution, for pH determination in the range of 8.0–11.2 (solution adjusted to 9.6) in Taylor Chemicals, Inc., comparator with Item 5166 slide.

5108   In 4-oz containers.

Filter paper, Whatman No. 5 quality, or equivalent, 12.5 cm, for filtration of boiler water before phosphate determination.

5109   Box of 100.

Filter paper, Whatman No. 1 quality, or equivalent, 12.5 cm, for filtration of softening or clarification plant water before analysis.

5110   Box of 100.

Standard sodium hydroxide solution, $N/50$ $(0.02\ N)$, for the titration of acidity.

5111   In plastic containers.

Mixed bromcresol green and methyl red indicator solution. Solution is prepared by dissolving 0.02 g methyl red and 0.1 g bromcresol green in 100 mL 95% ethyl alcohol or isopropyl alcohol. This indicator solution is used as the mixed indicator for the total alkalinity test.

5112   In pint containers.

5112A	In 8-oz containers.

Chlorphenol red indicator solution, for pH determination in the range of 5.2–6.8 (pH adjusted to 6.0) in Taylor Chemicals, Inc., comparator (such as Taylor Code #1003G).

5116	In 4-oz containers.

Sodium thiosulfate solution for determining chromate, $N/40$, stabilized, 1 mL = 20 ppm chromate (as CrO_4) when titrating a 50-mL water sample

5119	In quart containers.

Tripurpose chromate indicator powder, containing crystalline sulfamic acid (99+% purity) and dessicated ACS-grade granular potassium iodide in proper proportions for providing accurate results when determining chromate by thiosulfate titration of water samples. Approximately 4 g of this indicator powder sealed in a moisture-proof polyethylene aluminum foil envelope is required for each determination. An 8-oz carton contains 60 envelopes.

5120	In 8-oz cartons.

Standard ceric sulfate solution for determining nitrite, 0.110 N, 1 mL = 100 ppm nitrite (as NO_2) when titrating a 25-mL water sample taken from heating and chilled water systems.

5122	In quart containers.

Nitrite indicator solution, to be used with the standard ceric sulfate solution (Item 5122) to indicate the titration endpoint. This solution shall contain 0.70 g ferrous sulfate ($FeSO_4.7H_2O$) and 1.50 g o-phenanthroline ($C_{12}H_8N_2.H_2O$) dissolved in distilled water and diluted to 100 mL.

5123	In 5-oz containers.

Dilute sodium sulfate solution for checking accuracy of conductivity electrodes, used for determining conductivity of return condensate. Solution shall be 0.00038 N Na_2SO_4 and shall test 50 µS/cm at 25°C.

5128	In 1-gal bottles.

Sodium sulfate solution for checking accuracy of conductivity electrodes, used for determining conductivity of distribution, boiler, and cooling tower waters. Solution shall be 0.0093 N Na_2SO_4 and shall test 1000 µS/cm at 25°C.

5129	In 1-gal bottles.

Cylinder, 50-mL graduated, polymethylpentene or equivalent of maximum transparency for easy volume measurement, guaranteed accuracy of 1%, in 1-mL subdivisions.

5150   Each.

Cylinder, 100-mL graduated, polymethylpentene or equivalent of maximum transparency for easy volume measurement, guaranteed accuracy of 1%, in 1-mL subdivisions.

5151   Each.

Casserole, porcelain, 140-mL capacity, 85-mm diameter, 45-mm height (such as Coors size #3).

5152   Each.

Stirring rods, plastic, approximately 5 mm in diameter and 150 mm in length.

5153   Package containing 1 dozen.

Dropping bottle, 1 oz including rubber bulb and screw cap, with calibrated 0.5-mL dropper, for indicator solutions and reagents.

5154   Each.

Burette, automatic, 10 mL, graduated to 0.1 mL, including 1 pint-sized polyethylene (or equivalent) bottle, glass bead valve, clamp, glass tubing, and rubber connections; to be used for analyzing water samples.

5155   Each.

Funnel, analytical, top ID 75 mm, stem length 75 mm, maximum stem OD 12 mm, for 12.5-cm filter paper; constructed of polyethylene or equivalent plastic material; to be used for filtering water samples.

5156   Each.

Flask, Erlenmeyer, 250-mL capacity, glass, wide, including #8 rubber stopper.

5157   Each.

Bottle, wash or dispensing, 16 oz, screw cap with dispensing tip drawn to fine point, constructed of polyethylene or equivalent plastic material, to be used for dispensing molybdate reagent or distilled water.

5158   Each.

Colorimetric slide comparator, complete, with 9 color standards in slide, etched-glass compartment, five 5-mL test tubes, and two vials of indicator or reagent solutions required for conducting tests for constituents designated below (as manufactured by Taylor Chemicals, Sparks, MD 21152):

5159 *High-phosphate determination,* 0–100 ppm PO_4, including comparator and slide (such as Taylor Code #1100, including slide #1100B)

5160 *pH Determination, chlorphenol red,* including comparator and slide (such as Taylor Code #1010 and slide #1000G).

5161 *pH Determination, bromthymol blue,* including comparator and slide (such as Taylor Code #1010 and slide #1000H).

5162 *pH Determination, phenol red,* including comparator and slide (such as Taylor Code #1010 and slide #1000J).

5164 *Chromate determination,* 100–600 ppm sodium chromate (as Na_2CrO_4), including comparator and slide (such as Taylor Code 1130, Model N, and slide #1130A).

5165 *Low-phosphate determination,* 0–25 ppm PO_4, including comparator and slide (such as Taylor Code #1105 and slide #1105B.

5166 *pH Determination, thymol red,* in intervals of 0.4 unit, pH in range of 8.0–11.2, including comparator and slide (such as Taylor Code #1010 with slide #1000T).

5167 *pH Determination, meta-cresol purple,* including comparator and slide (such as Taylor Code #1010 and slide #1000L).

pH Slide, 6.0 to 7.6 range, bromthymol blue (such as Taylor Code #1000H).

5172 Each.

pH Slide, 6.8 to 8.4 range, phenol red (such as Taylor Code #1000J.)

5173 Each.

Test tubes, 5 mL, for Taylor Chemicals, Inc., comparators (such as Taylor Code #500).

5180 Each.

Reagent vials, 20-mL capacity, high-density polyethylene, with screw cap and bulb, 0.5-mL pipette, for Taylor Chemicals, Inc., comparators (such as Taylor Code #519A).

5181   Each.

Dalite lamp, midget, for Taylor Chemicals, Inc., comparators, 10 1/2 in long, 7 1/2 in wide, 4 1/2 in high, with Dalite glass, special bulb, cord, plug, 110 V ac, including spare lamp (such as Taylor Code #1070).

5182   Each.

Mixing tube for determining phosphate in boiler water, graduated at 5, 15, and 17.5 mL; 10-mm bottom diameter, 20-mm top diameter; including rubber stopper; for use in Taylor Chemicals, Inc., comparator (such as Taylor Code #558).

5183   Each.

Mixing tube for determining low phosphate in distribution and cooling waters, not for boiler water testing; graduated at 10 and 14 mL, 10-mm bottom diameter, 20-mm top diameter; for use in Taylor Chemicals, Inc., comparator (such as Taylor Code #559).

5184   Each.

Conductivity tester (boiler feedwater kit). Includes one Solubridge SD-152, one CEL-PS2 conductivity cell, 0 to 110°C thermometer (P/N5105-NE), 0.1-g dipper for gallic acid (P/N5101M), 50-mL graduated cylinder (P/N5104NE), 1 lb gallic acid, one SS-3 calibration solution to complete necessities for determining boiler water conductivity (as supplied by Rosemount Analytical Corp., 89 Commerce Road, Cedar Grove, NJ 07009, or equivalent).

5188   Each.

Conductivity tester for determining boiler water and condensate dissolved solids, dual range: 1–500 and 20–10,000 µS/cm, temperature compensator range 0–100°C; including a conductivity dip cell of 0.01 cell constant and one of 2.0 cell constant, and a switch for changing from one range to the other (such as SD125IF, manufactured by Rosemount Analytical Corp., 89 Commerce Road, Cedar Grove, NJ 07009, or equivalent).

5190   Each.

Conductivity dip cell, for condensate testing, cell constant 0.01, for determining conductivity of condensate by conductivity tester (Item

	5190) (such as Model CEL-PS001, manufactured by Rosemount Analytical Corp., 89 Commerce Road, Cedar Grove, NJ 07009, or equivalent).
5191	Each.

Conductivity dip cell, for boiler water testing, cell constant 2.0, for determining conductivity of boiler water by conductivity tester (Item 5190) (such as Model CEL-PS2, manufactured by Rosemount Analytical Corp., 89 Commerce Road, Cedar Grove, NJ 07009, or equivalent).

5192 Each.

Ungraduated cylinder for holding conductivity cells while testing, footed, height at least 5 in, ID at least 1 1/2 in, Plexiglas or equivalent.

5193 Each

Thermometer, dial type, with dial reset device, all metal, stainless-steel-alloy stem, 0 to 110°C range, for determining temperature in conductivity test.

5194 Each.

Brass or plastic 0.1-, 0.15-, and 1-g measuring cups, with metal handle, *as ordered for specifically measuring phosphate, hardness, calcium, and sulfite indicator powders.*

5197 Each, ordered for specific uses.

Plastic measuring cup, lucite cup with stainless-steel rod as handle, to hold 1 g.

5198 Each.

Cylinder, 10-mL graduated, glass, accuracy guaranteed within 1%, 0.2-mL subdivisions.

5199 Each.

Salometer for testing strength of salt solutions, hydrometer 0–100% [100% is maximum amount of salt (26.3% by weight) that can be dissolved by water at room temperature].

5200 Each.

Test-tube cleaning brush for Taylor Chemicals, Inc., 5-mL test tubes (such as Taylor Code #500).

5201 Each.

Chlorine comparator for determining free and total chlorine, by DPD method using individually packaged tablets; complete with nonfading comparator and tubes (such as Orbeco-Hellige Model 605HT with appropriate disc, 0.1–2.0 mg/L chlorine, or LaMotte Chemical Model 6817/LP-26, 6-octet comparator kit, 0.2–3.0 mg/L chlorine; or equivalent). (Orbeco-Hellige, 185 Marine Street, Farmingdale, NY 11735; LaMotte Chemical Products Co., P.O. Box 329, Chestertown, MD 21620.)

5205A   Each.

Tablets for determining free available chlorine by DPD method, for use with Orbeco-Hellige Model 605HT or equivalent. (Orbeco-Hellige, 185 Marine Street, Farmingdale, NY 11735.)

5205B   In packages of 50 or 250 as specified.

Tablets for determining total chlorine by DPD method, for use with Orbeco-Hellige Model 605HT or equivalent. (Orbeco-Hellige, 185 Marine Street, Farmingdale, NY 11735.)

5205D   In packages of 50 or 250 as specified.

Organophosphonate test kit, persulfate/ultraviolet light oxidation method, for the determination of phosphonate in cooling waters by conversion to orthophosphate. Includes light shield, power supply unit (120 V, 60 Hz, 36 W), safety goggles, comparator, and reagents (such as Cat. No. 22440-00 by Hach Co., P.O. Box 389, Loveland, CO 80539, or equivalent).

L5233   Each.

Direct reading spectrophotometer (such as Hach Co. Model DR/2000, Cat. No. 44800-00, (P.O. Box 389, Loveland, CO 80539), for analysis of some 120 water components.

L5234   Each.

Standard 5.25 N sulfuric acid solution, for adding to solution to be boiled in determining polyphosphate.

L5235   In pint containers.

Standard 5.0 N sodium hydroxide solution, for adding to solution after boiling in determining polyphosphate.

L5236   In pint containers.

Molybdate field test method, thioglycolate, comparator and axial reader, for determining molybdenum in the 2–20 mg/L range (such as Code/Model 3160/MBD by LaMotte Chemical Products Co., P.O. Box 329, Chestertown, MD 21620, or equivalent).

L5237   Each.

Zinc field test method, buffered zincon, comparator included, for determining zinc in the 0–10 mg/L range (such as Code/Model 7391/ZN by LaMotte Chemical Products Co., P.O. Box 329, Chestertown, MD 21620, or equivalent).

L5238   Each.

Glass electrode pH meter, complete (such as Hach Co. Cat. No. 43800-00, (P.O. Box 389, Loveland, CO 80539).

L5239   Each.

Bio-fouling monitor (such as Buckman Bio-Fouling Monitor, Buckman Laboratories, Inc., P.O. Box 080305, Memphis, TN 38108).

L5240   Each.

Sulfate-reducing bacteria test (SRB test) (such as Hach Co. No. 24324, P.O. Box 389, Loveland, CO 80539).

L5241   Package of 6.

Total bacterial count dip slides for determining total bacterial count in cooling waters, such as Easicult-TTC, Orion, including model chart for estimating total count after 48-h incubation and comparing density and characteristics of colonies on slide. (Available from Metalworking Chemicals & Equipment Co., 34 Main Street, P.O. Box 990, Lake Placid, NY 12946.)

L5242   Package of 10.

Appendix C

Sampling and Methods of Water Analysis

C.1 Sampling

Water tests are of no value unless the sample tested is representative of the water being analyzed. In fact, a nonrepresentative sample may indicate the need for a change in water treatment that may be unnecessary and could cause undesirable results.

Sample containers. Glass or plastic (polyethylene or polypropylene) is usually suitable for raw, distribution, cooling tower, and closed system water samples, as well as for boiler, feedwater, and condensate samples. Plastic bottles should be used if accurate silica results are to be obtained. Boiler water, feedwater, and return condensate should be sampled in glass if analyses for oil content are required. Samples should be stoppered immediately after collection.

Sampling from wells. It is essential that the well be pumped long enough (usually at least 15 min) in order to obtain a representative groundwater sample.

Sampling from lakes and reservoirs. Obtaining a representative sample may prove difficult; however, guidelines include avoiding sampling too close to the shore, too close to incoming flows, and in sluggish areas.

Sampling from a faucet. Water should be allowed to flow long enough so that the temperature of the water is fairly constant and approaches the temperature of the water in the larger piping. Also, the flow from the faucet should be regulated so that there is minimum splashing and aeration of the sample.

Sampling boiler water. Sampling of the continuous blowdown line flow and passage through a cooling coil is considered the proper sampling method, but the procedure below must be followed to attain this optimum method:

1. First, open the cooling water valve of the sample cooler to obtain necessary cooling.
2. Open the boiler water sampling valve for 15 to 30 s to blow sample lines free of stagnant water and sediment. A continuous flow of about 1 pt/min, cooled to below 120°F (50°C), should be maintained before the sample is collected.
3. Rinse the sample container at least 3 times with a small quantity of boiler water, shaking and then emptying.
4. Take the sample when the boiler is operating under normal load conditions at normal boiler water level and before manual blowdowns have been applied. Water test results will be expected to indicate the amount of needed manual blowdown or the required adjustment of the continuous blowdown valve.
5. Collect the boiler water sample through a pipe extending to the bottom of the container. Allow the container to overflow at least 5 volumes; then close the container with a stopper. Boiler water samples containing sulfite will be expected to pick up oxygen from the air if the containers are not closed.
6. After the sample is taken, shut off the boiler water sampling valve and then the cooling water valve of the sample cooler.

Sampling return condensate. Use the following procedure:

1. Sample the total condensate flow in order to obtain a more complete test of the overall condensate purity and freedom from raw or softened water contamination from heat exchangers in the system.
2. Sample condensate in a glass container (labeled "Only for condensate"), and cool the sample to room temperature before testing. Sample containers that previously contained boiler or raw water may retain sufficient dissolved solids in seams to contaminate condensate samples.
3. Rinse the sample container at least 3 times with condensate before retaining the sample to be tested.

4. Test the sample immediately to avoid contamination from air or dust. Do not filter the sample.

Sampling from low-pressure boilers or those not having continuous blowdown. Sampling must therefore be done from the water column. In this case the water column should be blown down at least twice to ensure that fresh water is obtained from the boiler. In the case of boilers that have automatic feedwater addition at the water column, the automatic device must be *temporarily shut off* during sampling in order to avoid contaminating the boiler water sample with feedwater. (Be sure to turn the automatic feedwater device back on after sampling.) Also care must be taken not to sample too fast from the water column, as condensed steam from above the water level may be sampled and will dilute the actual boiler water sample.

C.2 Water Testing and Methods of Water Analysis

The only purpose of water testing is to provide proper control of recommended treatments. It is *not* for the purpose of recordkeeping or of completing forms. When tests indicate undesired results, adjustment of treatment, blowdown, etc., as recommended in the control charts should be immediately applied. A minimum number of tests are prescribed, and the results are essential in maintaining physical plant facilities in the desired operating and maintenance-free condition.

Testing equipment and replacement items should be ordered according to App. B. Reagents in general will maintain the specified concentration for at least 1 year. Occasionally, solutions from a particular bottle or supply are off-standard. When test results change radically with use of a new bottle of reagent, it is possible that the new supply may be off-standard. A sample of the off-standard reagent can be sent to your consultant for checking.

Testing equipment should be kept scrupulously clean and in good working order. All bottles should be properly labeled. Spilled chemicals should be cleaned immediately from tabletops. Graduated cylinders and casseroles should be rinsed with the water to be tested before the analysis is conducted. Care should be taken in measuring the test volume of the water sample, in reading the automatic burettes, and in reporting the mg/L (ppm) results.

The following methods in C.2 are mainly analytical methods using burettes for titration of water quality items.

C.2.1 Hardness test (H)

Apparatus

Burette, 10-ml automatic (for standard hardness titrating solution) (Item 5155)

Graduated cylinder (Item 5150)

1-oz bottle, with calibrated 0.5-mL dropper (Item 5154)

Porcelain casserole (Item 5152)

Stirring rod (Item 5153)

Brass measuring cup, 0.1-g size (Item 5197)

Reagents

Standard hardness titrating solution (1 mL = 1 mg $CaCO_3$) (Item 5002 or 5003)

Hardness buffer solution (Item 5011)

Hardness indicator powder (Item 5014)

Method

1. Measure the amount of water to be tested in the graduated cylinder in accordance with the following:

Hardness (as $CaCO_3$), mg/L	Recommended sample size (to keep titration less than 10 mL), mL	Factor
Up to 100	100	10
100 to 400	50	20
Above 400	20	50
	10	100
	5	200

2. Pour into casserole.
3. Add 1 mL hardness buffer and 1 brass measuring cup (0.15 g) of hardness indicator powder. If the water turns blue, no hardness is present, the hardness is reported as "zero," and the test is discontinued. If the water turns red, hardness is present and the test is continued.
4. Squeeze the standard hardness titrating solution from the plastic

bottle to just above the zero mark on the burette; then allow excess solution to drain automatically back into the bottle.

5. While stirring the water constantly with the glass rod, add standard hardness titrating solution drop by drop from the burette until the red color changes to blue. This is the endpoint. Read the burette to the nearest 0.1 mL.

Results

When testing a 100-mL sample,

$$\text{Burette reading in mL} \times 10 = \text{mg/L hardness (as CaCo}_3\text{)}$$

Example:

$$5.6 \text{ mL} \times 10 = 56 \text{ mg/L hardness (as CaCo}_3\text{)}$$

C.2.2 Calcium (Ca) and magnesium (Mg) hardness tests

Apparatus

Burette, 10-mL automatic (for standard hardness titrating solution) (Item 5155)

Burette, 10-mL automatic (for standard sulfuric acid solution, $N/50$) (Item 5155)

Graduated cylinder (Item 5150)

Porcelain casserole (Item 5152)

Stirring rod (Item 5153)

Bottle, with dropper (for methyl orange indicator solution) (Item 5154)

1-oz bottle, with calibrated 0.5-mL dropper (for sodium hydroxide) (Item 5154)

Brass measuring cup, 0.1-g size (Item 5197)

Reagents

Standard sulfuric acid solution ($N/50$) (Item 5021 or 5022)

Methyl orange indicator solution (Item 5029)

Standard hardness titrating solution (1 mL = 1 mg $CaCO_3$) (Item

5002 or 5003)

Sodium hydroxide solution (1.0 N) (Item 5095)

Calcium indicator powder (Item 5099)

Method

1. Measure the amount of water to be tested in the graduated cylinder in accordance with the following:

Hardness (as $CaCO_3$), mg/L	Recommended sample size (to keep titration less than 10 mL), mL	Factor
Up to 100	100	10
100 to 400	50	20
Above 400	20	50
	10	100
	5	200

2. Pour into casserole.

3. Squeeze the plastic bottles containing the standard sulfuric acid solution and the standard hardness titrating solution to provide solution just above the zero marks on the burettes; then allow the excess solutions to drain automatically back into the bottles.

4. Add 2 to 3 drops of methyl orange indicator solution, and titrate with N/50 sulfuric acid solution until the color changes from yellow to orange or faint pink.

5. Now apply 2 brass measuring cups of calcium indicator powder and 2 mL (1 N) sodium hydroxide solution.

6. Then titrate immediately with hardness titrating solution. If calcium is present, the sample will be salmon pink in color, and as the endpoint is approached, it will develop a purple tinge. The endpoint is orchid purple. Once the endpoint is reached, additional titrating solution *will not produce any further color change*. The endpoint should always be checked by adding an additional drop of titrating solution and observ-

ing whether further color change occurs. If further color change occurs, the endpoint has not been reached and the titration should be continued. Read the burette to the nearest 0.1 mL.

Results

Calcium hardness (as $CaCO_3$). When testing a 100-mL sample,

Burette reading in mL × 10 = mg/L calcium hardness (as $CaCO_3$)

Example:

3.6 mL × 10 = 36 mg/L calcium hardness (as $CaCO_3$)

Magnesium hardness (as $CaCO_3$)

Total hardness (mg/L) − calcium hardness (mg/L)
   = magnesium hardness (mg/L as $CaCO_3$)

C.2.3   Phenolphthalein alkalinity test (P)

Apparatus

Burette, 10-mL automatic (for standard sulfuric acid solution) (Item 5155)

Graduated cylinder (Item 5150)

Bottle, with dropper (for phenolphthalein indicator solution) (Item 5154)

Porcelain casserole (Item 5152)

Stirring rod (Item 5153)

Reagents

Standard sulfuric acid solution ($N/50$) (1 mL = 1 mg $CaCO_3$) (Item 5021 or 5022)

Phenolphthalein indicator solution (Item 5033)

Method

1. Measure the amount of water to be tested in the graduated cylinder in accordance with the following:

P alkalinity (as CaCO$_3$), mg/L	Recommended sample size, mL	Factor
Up to 25	100	10
25–100	50	20
100–400	20	50
Above 400	10	100

2. Pour into casserole.
3. Add 2 drops of phenolphthalein indicator solution. If the water does not change to a pink color, there is no phenolphthalein alkalinity present and the P alkalinity is reported as zero. If the water changes to a pink color, phenolphthalein alkalinity is present and the test should be continued.
4. Squeeze the standard sulfuric acid solution from the plastic bottle to just above the zero mark on the burette, and then allow the excess acid to flow automatically back into the bottle.
5. While stirring the sample constantly with the stirring rod, add standard sulfuric acid solution drop by drop from the burette until the pink color disappears and the water has the color it showed before the phenolphthalein indicator was added. This is the endpoint. Read the burette to the nearest 0.1 mL.
6. Do not discard the sample; save it if M alkalinity and/or chloride tests are required.

Results

When testing a 100-mL sample:

Burette reading in mL × 10 = mg/L P alkalinity (as CaCO$_3$)

Example:

7.0 mL × 10 = 70 mg/L P alkalinity (as CaCO$_3$)

C.2.4 Total (methyl orange) alkalinity test (M)

Apparatus

Burette, 10-ml automatic (for standard sulfuric acid solution) (Item 5155)

Graduated cylinder (Item 5150)

Bottle, with dropper (for methyl orange indicator solution) (Item 5154)

Porcelain casserole (Item 5152)

Stirring rod (Item 5153)

Reagents

Standard sulfuric acid solution ($N/50$) (1 mL = 1 mg $CaCO_3$) (Item 5021 or 5022)

Methyl orange indicator solution (Item 5029)

Method

1. Add 2 to 4 drops of methyl orange indicator solution to the water in the casserole remaining from the phenolphthalein alkalinity test. If the water changes to an orange or red color, free mineral acid may be present, there is no alkalinity, and the alkalinity is reported as zero. If the water changes to a yellow or straw color, methyl orange alkalinity is present and the test should be continued.
2. Do not squeeze the standard sulfuric acid solution from the plastic bottle back to the zero mark on the burette, but leave it at the reading shown at the endpoint of the P alkalinity test.
3. While stirring the water constantly with the stirring rod, add standard sulfuric acid solution drop by drop from the burette until the yellow or straw color changes to a faint orange or light pink. Note the total burette reading to the nearest 0.1 mL, including the reading in the P alkalinity determination.
4. Do not discard the sample if a chloride test is required.

Results

When testing a 100-mL sample:

Total burette reading in mL × 10 = mg/L M alkalinity (as $CaCO_3$)

Example:

7.0 mL × 10 = 70 mg/L P alkalinity (as $CaCo_3$)

10.0 mL × 10 = 100 mg/L M alkalinity (as $CaCO_3$)

Note: The endpoint may be difficult to see in artificial light and can be determined more accurately in daylight. Inexperienced operators generally overtitrate the endpoint, obtaining high results and a definitely red endpoint. There is a mixed indicator (Item 5112) that may be used in case of a real problem in discerning the methyl orange endpoint.

C.2.5 Alkalinity in condensate

The purpose of this test is to determine the approximate amine content so that the proper control of amine treatment may be maintained.

After the condensate sample is collected as directed in Sec. C.1, the P and M alkalinity tests are conducted.

Apparatus

Burette, 10-mL automatic (for standard sulfuric acid solution) (Item 5155)

Graduated cylinder, 100 mL (Item 5151)

Two bottles, with droppers (for phenolphthalein indicator solution and methyl orange indicator solution (Item 5154)

Porcelain casserole (Item 5152)

Stirring rod (Item 5153)

Reagents

Standard sulfuric acid solution ($N/50$) (1 mL = 1 mg $CaCO_3$) (Item 5021 or 5022)

Phenolphthalein indicator solution (Item 5033)

Methyl orange indicator solution (Item 5029)

Method

1. Mark a 100-mL graduated cylinder and a casserole to be used for *condensate analytical purposes only.*
2. After rinsing the cylinder and casserole with the condensate sample several times, measure a 100-mL sample and add it to the casserole.
3. Add 2 drops of phenolphthalein indicator solution. If the sample remains colorless, record P alkalinity as zero.
4. If the sample is pink, add $N/50$ sulfuric acid solution until the sample is colorless. This is the endpoint. Read the burette to the nearest 0.1 mL.

5. Do not squeeze the standard sulfuric acid solution from the plastic bottle back to the zero mark on the burette, but leave it at the reading shown at the endpoint of the P alkalinity test.
6. Now add 3 to 4 drops of methyl orange indicator solution and continue titration with $N/50$ sulfuric acid solution to the pink endpoint. Read the burette to the nearest 0.1 mL.

Results

Burette reading to phenolphthalein endpoint × 10 =
$\qquad$ mg/L P alkalinity (as $CaCO_3$)

Total burette reading to methyl orange endpoint × 10 =
$\qquad$ mg/L M alkalinity (as $CaCO_3$)

Example:

0.8 mL required to phenolphthalein endpoint × 10 = 8 mg/L P alkalinity

1.6 mL required to methyl orange endpoint × 10 = 16 mg/L M alkalinity

Interpretation

Usually it is recommended that the P alkalinity be maintained at a slight positive value, but below 6 mg/L, and that the M alkalinity be maintained below 30 mg/L.

C.2.6 pH

The pH measurement using the glass electrode is the most accurate method of determining pH; however, this equipment is generally expensive and is difficult to maintain. The glass electrode technique is described in Sec. C.3.

There are also colorimetric pH methods, in which colorimetric comparators and various indicators, such as chlorphenol red (Item 5116), bromthymol blue (Item 5107), phenol red (Item 5106), thymol red (Item 5108), and *meta*-cresol purple (Item 5167), are used. These methods are not recommended for testing the pH of return condensate, as the indicator solutions themselves are so much stronger in pH than condensate that the results reflect the indicator pH and not the condensate pH. These pH methods may be used in their particular pH ranges for testing waters of higher conductivity than return condensate.

C.2.7 pH Estimation from P and M alkalinity readings

Method

1. Determine P alkalinity and M alkalinity by Methods C.2.3 and C.2.4.
2. Then refer to Fig. C.1. Note the P alkalinity in the horizontal scale; then move upward on the vertical scale to the M (methyl orange) alkalinity.
3. Note the position of the point on the graph; note the position relative to the slanting lines (pH curves) and estimate the pH.

Example 1	Example 2
P alkalinity = 30 mg/L	P alkalinity = 49 mg/L
M alkalinity = 120 mg/L	M alkalinity = 150 mg/L
Estimated pH = 9.8	Estimated pH = 10.1

4. Or, determine pH by dividing the M alkalinity by the P alkalinity and referring to the following table:

M/P	pH	M/P	pH	M/P	pH
60	8.5	7	9.5	3.6	10.0
18	9.0	5.8	9.6	2.9	10.2
10	9.3	4.2	9.8	2.4	10.4

C.2.8 Chloride test (Cl)

Apparatus

Burette, 10-mL automatic (for standard silver nitrate solution) (Item 5155)

Graduated cylinder (Item 5150)

Bottle, with droppers (for potassium chromate indicator solution) (Item 5154)

Porcelain casserole (Item 5152)

Stirring rod (Item 5153)

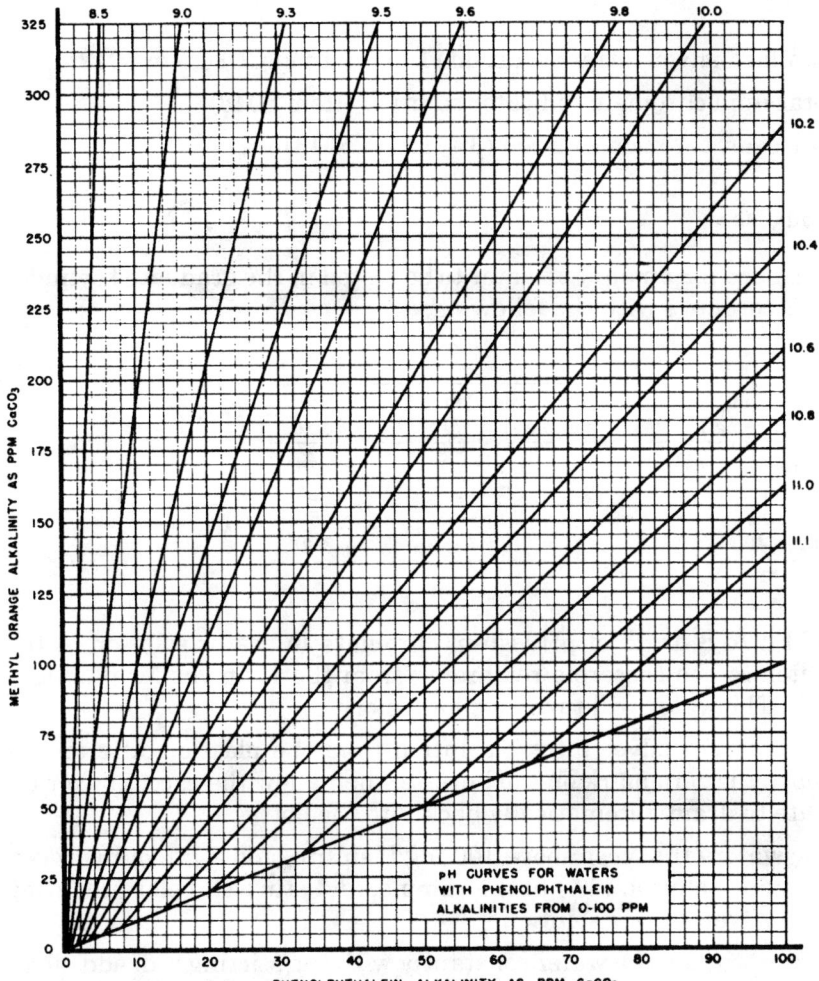

NOTE: pH value will also depend on temperature of water. Chart above is based on temperature of 20 to 25°C. As water temperature decreases, the pH value for any given combination of alkalinity forms will increase slightly above the value indicated on the chart. For example, at 5°C. actual pH will be about 0.2 units higher in 8.5 to 9.0 pH range; about 0.3 units higher in 9.0 to 10.0 pH range; and above pH 10 actual pH will be 0.4 to 0.6 pH units higher than indicated by chart.

Figure C.1 Determination of pH from P and M alkalinity. (*Permutit Water Conditioning Handbook, 1954; now Water and Waste Treatment Databook, The Permutit Co., 30 Technology Drive, Warren, NJ 07059.*)

Reagents

Standard silver nitrate solution (1 mL = 1 mg Cl⁻) (Item 5042)
Potassium chromate indicator solution (Item 5048)
Hydrogen peroxide solution (3%) (Item 5043)

Method

1. Measure the amount of water to be tested in the graduated cylinder in accordance with the following:

Chloride (as Cl⁻), mg/L	Recommended sample size, mL	Factor
Up to 25	100	10
25–100	50	20
100–400	20	50
Above 400	10	100

2. Add 4 to 8 drops of potassium chromate indicator solution to the water in the casserole remaining from the P and/or M alkalinity test. The water will turn a yellow color.

3. Squeeze the silver nitrate solution from the plastic bottle to just above the zero mark on the burette; then allow the excess silver nitrate to drain automatically back into the bottle.

4. For waters that test above 100 mg/L sulfite, add 2 mL of hydrogen peroxide solution (3%) to the sample, and stir well before titrating with silver nitrate solution.

5. While stirring the water constantly with the stirring rod, add silver nitrate solution drop by drop from the burette until a faint red or dark orange color appears *throughout* the entire sample. Read the burette to the nearest 0.1 mL.

Results

When testing a 100-mL sample:

Burette reading in mL × 10 = mg/L chloride (as Cl⁻)

Note A: The endpoint is the *first* appearance of a permanent faint red or dark orange color and *not* the deep red color that develops if silver nitrate is added past the endpoint.

Note B: The P alkalinity test *must* be conducted first, before the chloride is determined.

C.2.9 Sulfite test (SO_3)

Apparatus

Burette, 10-mL automatic (for standard potassium iodide-iodate solution) (Item 5155)

Graduated cylinder (Item 5150)

1-oz bottle, with droppers (for phenolphthalein indicator solution (Item 5154)

Porcelain casserole (Item 5152)

Stirring rod (Item 5153)

1 Plastic dipper (for 1.0 g dual-purpose starch indicator powder (Item 5198)

Reagents

Standard potassium iodide-iodate solution (1 mL = 1.0 mg SO_3^{2-}) (Item 5068)

Phenolphthalein indicator solution (Item 5033)

Dual-purpose sulfite indicator powder (Item 5072)

Method

1. Add 1 plastic dipper (1.0 g) of dual-purpose sulfite indicator powder to casserole.
2. Measure the amount of water to be tested in the graduated cylinder in accordance with the following:

Sulfite (SO_3)	Recommended sample size, mL	Factor
Up to 100	50	20
100–500	10	100

3. Pour sample into casserole, stir well, and add 3 drops of phenolphthalein indicator solution.
4. If solution turns red, add more dual-purpose sulfite indicator powder until solution returns to its original color.
5. Squeeze the standard potassium iodide-iodate solution from the plastic bottle to just above the zero mark on the burette; then allow the excess solution to drain automatically back into the bottle.

6. While stirring the sample constantly, add the potassium iodide-iodate solution drop by drop until a faint but permanent blue color appears. This is the endpoint. Be careful not to overtitrate the endpoint. Read the burette to the nearest 0.1 mL.

Results
When testing a 50-mL sample:

$$\text{Burette reading in mL} \times 20 = \text{mg/L sulfite (as SO}_3)$$

Example:

$$1.4 \text{ mL} \times 20 = 28 \text{ mg/L sulfite (as SO}_3)$$

Note: To attain the most accurate results, the sample should be collected without contact with the air, be cooled to below 120°F (50°C), and tested promptly.

C.2.10 Phosphate test (PO_4)

Apparatus

Taylor high-phosphate comparator, complete, range 0 to 100 mg/L (ppm) (Item 5159)

High-phosphate mixing tube graduated at 5, 15, and 17.5 mL, with rubber stopper (Item 5183)

2 Test tubes, 5 mL (Item 5180)

Dispensing bottle, pint size (for molybdate reagent solution) (Item 5158)

Funnel (Item 5156)

Filter paper, Whatman No. 5, 12.5-cm diameter (Item 5109)

Taylor Dalite lamp, 110 V, 60 cycles (Item 5182)

Brass measuring cup, 0.1-g size (Item 5197)

Test-tube cleaning brush (Item 5201)

Reagents

Phosphate indicator powder (Item 5075)

Molybdate reagent solution (Item 5084)*

Distilled water

---

*Contains strong acid and should be handled with caution.

Preparation of water sample to be tested

It is imperative that the water sample to be tested be free of suspended matter or sludge. The sample shall be filtered through the same filter paper as many times as necessary to produce a clear sample. A new filter paper shall be used to filter each new sample.

Method

1. Fill the phosphate mixing tube to the 5-mL mark with the filtered water sample.
2. Add molybdate reagent solution to the 15-mL (middle) mark.
3. Add 2 measuring cups full of phosphate indicator powder to the mixing tube; add distilled water to the 17.5-mL mark; insert the rubber stopper; and mix.
4. If a blue color does not form in 5 min, phosphates are absent and should be reported as zero. If a blue color develops, place the mixing tube in the middle hole of the phosphate comparator.
5. Fill the two 5-mL test tubes with filtered boiler water and place in the holes on each side of the mixing tube.
6. Place the comparator on the Dalite lamp shelf. Turn the switch on, and move the comparator slide until the intensity of the color of the sample is matched by one of the color standards of the comparator.

Results

The phosphate (PO_4) in milligrams per liter (parts per million) is the number appearing on the slide as indicated by the arrow on the base. If the phosphate exceeds 100 mg/L (ppm), discard this test and repeat the test using a 2.5-mL sample diluted with 2.5 mL of distilled water in the 5-mL portion of the mixing tube. On completion of the test, multiply the reading by 2 to obtain the mg/L (ppm) phosphate.

C.2.11 Nitrite test (NO_2) (ceric sulfate method)

Apparatus

Burette, 10-mL automatic (for 0.110 N ceric sulfate solution) (Item 5155)

Graduated cylinder (Item 5150)

1-oz bottle, with dropper (for nitrite indicator solution) (Item 5154)

Porcelain casserole (Item 5152)
Stirring rod (Item 5153)

Reagents

Standard ceric sulfate solution (0.110 N) (Item 5122)
Nitrite indicator solution (Item 5123)

Method

1. Measure 25 mL in graduated cylinder and pour into casserole.
2. Add exactly 1 full drop of nitrite indicator solution to the water sample, which will turn red if nitrite is present.
3. Squeeze the ceric sulfate solution from the plastic bottle to just above the zero mark on the burette; then allow the excess ceric sulfate to flow automatically back into the bottle.
4. While stirring the sample vigorously with the stirring rod, add ceric sulfate solution drop by drop until the red color changes to blue and *remains blue for a full 15 s*. This is the endpoint. Read the burette to the nearest 0.1 mL.

Results

When testing a 25-mL sample:

$$\text{Burette reading in mL} \times 100 = \text{mg/L nitrite (as NO}_2\text{)}$$

Example:

$$5.6 \text{ mL} \times 100 = 560 \text{ mg/L nitrite (as NO}_2\text{)}$$

C.2.12 Chromate test (CrO$_4$)

Apparatus

Burette, 10-mL automatic (for $N/40$ sodium thiosulfate solution) (Item 5155)
Graduated cylinder (Item 5150)
Porcelain casserole (Item 5152)
Stirring rod (Item 5153)

Reagents

N/40 sodium thiosulfate solution (Item 5119)
Tripurpose chromate indicator powder, envelopes (Item 5120)
Distilled water

Method

1. Measure the amount of water to be tested in the graduated cylinder in accordance with the following:

Chromate (as CrO_4), mg/L	Recommended sample size, mL	Factor
0–100	50	20
Above 100	20	50

Dilute the 20-mL sample to 50 mL with distilled water in the graduated cylinder before pouring into the casserole.

2. Open one envelope of tripurpose chromate indicator powder, pour entire contents into casserole, and stir with stirring rod until indicator is dissolved.
3. Squeeze the sodium thiosulfate solution from the plastic bottle to just above the zero mark on the burette; then allow excess solution to drain automatically back into the bottle.
4. Titrate with N/40 sodium thiosulfate solution until water sample turns to a straw yellow. At this point the endpoint is near and titration should be conducted slowly. The endpoint is when the yellow color disappears and the sample turns to a clear or light blue color. Read the burette to the nearest 0.1 mL.

Results

When testing a 50-mL sample:

$$\text{Burette reading in mL} \times 20 = \text{mg/L chromate (as } CrO_4)$$

Example:

$$2.0 \text{ mL} \times 20 = 40 \text{ mg/L chromate (as } CrO_4)$$

Note: In the presence of other oxidizing agents, this procedure may yield high results for chromate. This may be checked by using the diphenylcarbazide colorimetric procedure (as provided by Hach Co. or others).

C.3 Instrumental Test Methods

These test methods use instruments such as an electrical conductivity meter, a glass electrode pH meter, and a spectrophotometer or colorimeter for determining different water quality factors or constituents that are important in providing proper control of scale and corrosion in water systems.

Temperature has a pronounced effect on conductivity measurements; measurement at 25°C ±0.1°C is considered to provide the most accurate results. Most conductivity measurements provide temperature compensation, but one should realize that such compensation must be based on a specific composition and is therefore subject to error when the composition is unknown.

On-line measurement of conductivity using flow-type cells provides a more accurate value than static measurement, as then gaseous and other contaminants are not present to provide erroneous values. Also, ions such as sodium and chloride are most accurately measured by sensor electrodes under these conditions.

Conductivity measurements may include specific conductivity, cation conductivity (for more sensitive detection or condenser leakage), and anion conductivity, which provides increased accuracy of chloride and sulfate contamination by eliminating errors introduced by the presence of carbon dioxide (Fig. 4.15). An estimation of the carbon dioxide (air in-leakage in a turbine condenser) may be attained by on-line conductivity measurements, as described in ASTM D4519 (see Chap. 4, Ref. 17).

The following methods for static and specific conductivity are simpler methods, which may include manual temperature adjustment or automatic temperature compensation. It is specified that the water to be tested should be cooled to at least 120°F (50°C) before measurement.

C.3.1 Electrical conductivity test (for boiler and cooling tower dissolved solids)

Apparatus

Electrical conductivity instrument for 105–130 V, 50–60 cycles, dual range: 1–500 and 20–10,000 µS/cm, temperature compensator range 0–100°C (Item 5188 or 5190), including conductivity dip cells having cell constants of 0.01 and 2.0 (Items 5191 and 5192).

ac Electrical outlet

Dial thermometer (Item 5194)

100-mL-capacity cylinder, not graduated (Item 5193)

Plastic measuring cup, 1.0-g size (for gallic acid) (Item 5197)

Reagents

Phenolphthalein indicator solution (Item 5033)

Gallic acid (Item 5057)

Method

1. Connect instrument to 100-V ac electrical outlet, connect dip cell, and turn instrument switch to "on" position.
2. Pour approximately 50 mL of sample into cylinder (sufficient to cover air vents of dip cell).
3. Add 2 to 4 drops of phenolphthalein indicator solution and sufficient gallic acid to change the sample solution from pink to colorless. [Usually 2 cupfuls (Item 5197) is sufficient.]
4. Determine temperature of sample with thermometer, and set instrument temperature dial at this temperature. If conductivity instrument includes automatic temperature compensation, it is unnecessary to take the temperature and to set temperature in the meter.
5. Set range at 20–10,000 µS/cm.
6. Now immerse the dip cell in the sample, and move the cell (cell constant 2.0) up and down several times under the water surface to remove air bubbles from inside the cell shield. (Keep the dip cell at least 1/2 in off the bottom of the cylinder when taking the conductivity reading.)
7. Now read the microsiemens per centimeter from the instrument dial; correct the reading according to the cell constant of 2.0 (multiply by 2). Record the reading.

Results

Report the results in microsiemens per centimeter, as read from the dial. Results in milligrams per liter dissolved solids may be approximated by multiplying the dial reading by a factor of 0.75 to 0.9, depending on the type of water sample being tested.

C.3.2 Electrical conductivity test (condensate dissolved solids)

The presence of mineral solids in condensate, caused by contamination of condensate by raw or softened water or by other contamination, may be determined by an electrical conductivity instrument having an accurate range of measurement of 0–50 µS/cm.

The detection of mineral solids in condensate is very important, since the presence of raw or softened water or other contamination in the condensate system may indicate serious boiler water carryover, or it can cause boiler carryover and overloading of the feedwater system, which will result in increased scale and corrosion and decreased efficiency of the entire power plant.

Apparatus

Electrical conductivity instrument for 105–130 V, 50–60 cycles, dual range: 1–500 and 20–10,000 µS/cm, temperature compensator range 0–100°C (Item 5190), including conductivity dip cells having cell constants of 0.01 and 2.0 (Items 5191 and 5192).

ac Electrical outlet

Dial thermometer (Item 5194)

100-mL-capacity cylinder, not graduated (Item 5193)

Reagents
None.

Method

1. Allow condensate to flow for several minutes before sampling (see Sec. C.1). Rinse container several times with the sample before collecting for analysis. Collect sample in a clean (preferably glass) container (labeled *for condensate only*); the sample should preferably be cooled by passage through a small heat exchanger.
2. Connect instrument to 110-V ac electrical outlet, connect dip cell, and turn instrument switch to "on."
3. Pour approximately 50 mL of sample into cylinder, rinse electrode and thermometer. Take temperature reading, and set temperature dial accordingly.
4. Dip cell (0.01 cell constant) should rest 1/2 in or more off the bottom of the cylinder when the conductivity reading is taken. Repeat rinsing until consistent reading is obtained. Correct the reading according to the cell constant of 0.01 (divide by 100).

Results

Report the results in microsiemens per centimeter, as read from the instrument. Results in milligrams per liter may be approximated by multiplying the instrument reading by 0.6.

Note: Values above 20–30 µS/cm may indicate that contamination from raw or softened water is occurring owing to a heat exchanger leak.

C.3.3 Test to check conductivity instrument and electrodes

The electrical conductivity instrument and electrodes should be checked weekly for accuracy by testing the conductivity of standard solutions (Items 5128 and 5129—sodium sulfate solutions).

Instructions for checking the boiler water electrode (cell constant 2.0) (Item 5129) are as follows:

1. Connect instrument to 110-V ac electrical outlet, turn on the instrument switch, connect dip cell, and place dip cell in ungraduated plastic cylinder (Item 5193).
2. Rinse electrode well, preferably with condensate, and shake excess water from cell.
3. Pour sufficient sodium sulfate solution (Item 5129) into cylinder to cover air vents of cell, usually about 50 mL. Temperature of sodium sulfate solution should be in the 60–100°F (15–40°C) range.
4. After immersing dip cell in the sample, move cell up and down several times under the water to remove air bubbles from inside the cell shield.
5. Determine the temperature of the sample with the thermometer (Item 5194), and set temperature dial accordingly on the instrument.
6. Keep the dip cell 1/2 in off the bottom of the cylinder when taking the conductivity reading from the instrument.
7. Report the result in microsiemens per centimeter weekly on the water treatment report. The reading should be 900–1100 µS/cm.

Instructions for checking the condensate electrode (0.01 cell constant) are similar to the above, but the weaker sodium sulfate solution (Item 5128) (45–55 µS/cm) should be used. Electrodes may be stored dry on completion of the testing.

Further instructions

1. In case results are not in the specified ranges, the electrode should be soaked in successive lots of $N/50$ sulfuric acid, or if a more rapid

cleaning procedure is preferred, it may be immersed for 1 min only in (1:1) hydrochloric acid, prepared from Item 40 hydrochloric acid or obtained from a chemical supply house. After the electrode is cleaned, it should again be tested in standard sodium sulfate solution. If the proper test result is then not obtained, it is likely that the electrode requires replatinization, repair, or replacement and should be returned to the supplier.

2. Spare conductivity electrodes (Items 5191 and 5192) should be kept on hand at all times. Since conductivity measurement is such an important and basic measurement for power plant water treatment control, this ensures that a means of accurate conductivity measurement is available at all times.

C.3.4 pH Measurement by glass electrode

In conducting the pH determination using a pH meter like that described in Item L5239, reference should be made to the test procedure described in the *Hach Water Analysis Handbook* (Hach Co., Loveland, Colo., 1989).

C.3.5 Photometric determination of various water constituents

Numerous field-type and laboratory instruments are available, but the author's experience in water treatment consulting work has revealed that the Hach DR/1A Colorimeter and DR/2000 Spectrophotometer (Item L5234) (Fig. C.2) have provided very satisfactory results. These

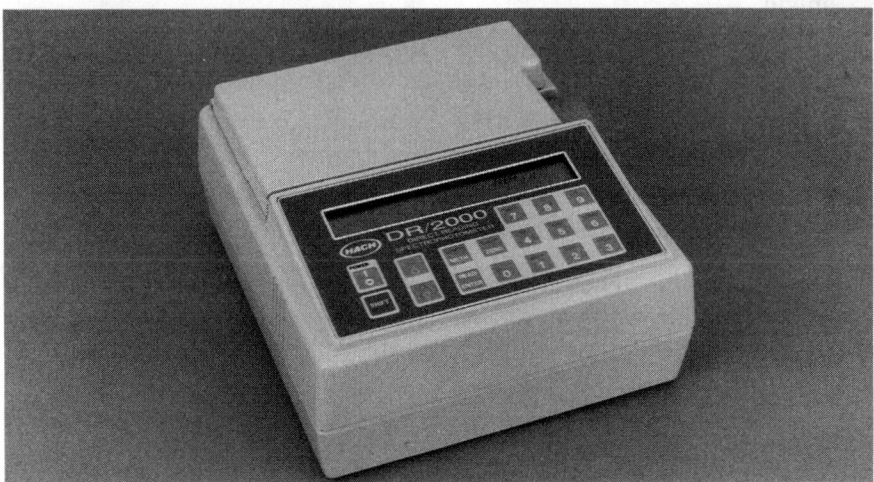

Figure C.2 Direct Reading Spectrophotometer programmed for conducting some 29 water quality tests. (*Hach Co., P.O. Box 839, Loveland, CO 80539.*)

instruments have been used for the determination of low and high levels of aluminum, bromine, calcium and magnesium (0–4.00 mg/L Ca and Mg as $CaCO_3$), chlorine (free and combined), hydrazine, iron, manganese, molybdenum, nitrate, orthophosphate, polyphosphate, phosphonate, silica, tannin (lignin), turbidity, and zinc in industrial waters. Apparatus for determining chemical oxygen demand (COD) has also been used for determining the amount of organic matter in cooling water (Fig. C.3). In making any of these determinations, reference is made to the test procedures in the *Hach Water Analysis Handbook* (Hach Co., Loveland, Colo., 1989). For determining chelants in boiler feedwater, refer to the *Hach Digital Titrator Methods Manual*.

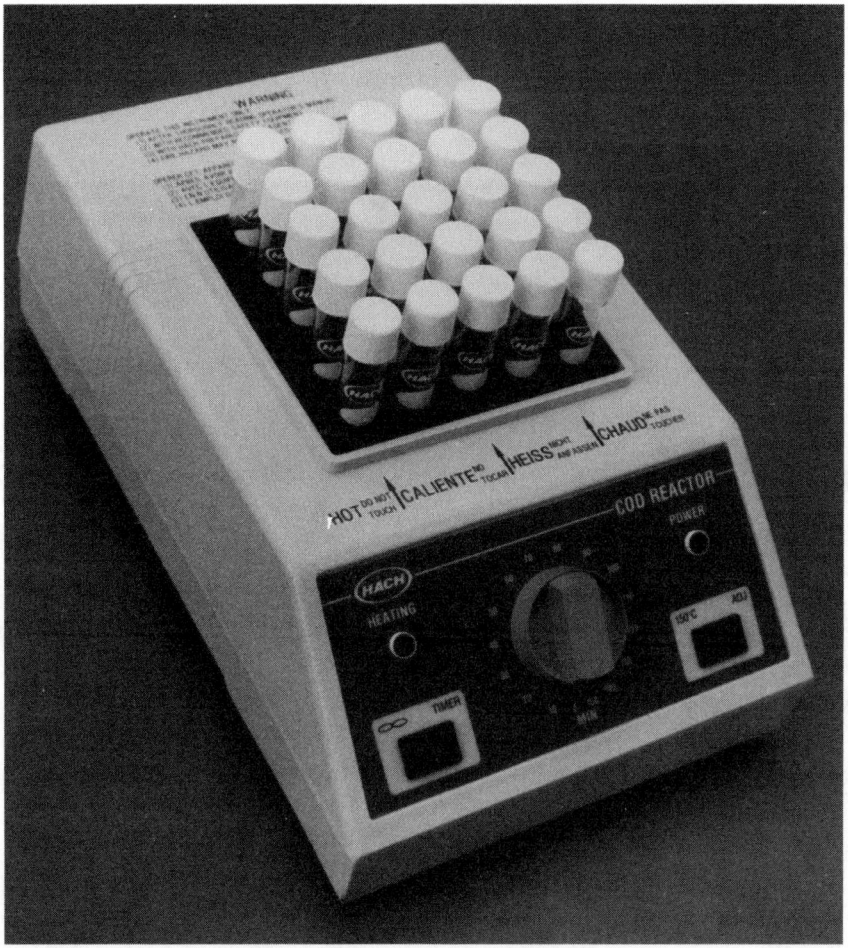

Figure C.3 COD Reactor for determining chemical oxygen demand using a dry bath and reagent vials. (*Hach Co., P.O. Box 839, Loveland, CO 80539.*)

C.3.6 Field-type colorimetric determinations of molybdate, zinc, and phosphonate

Methods mainly used for field-testing and listed as Items L5237, L5238, and L5233 have been found satisfactory. In conducting these determinations, reference should be made to the test procedure provided by the manufacturer of the test equipment.

C.3.7 Field-type tests for total bacterial count and sulfate-reducing bacteria in cooling waters

Test methods listed as Items L5241 and L5242 have been satisfactory. In conducting these determinations, reference should be made to the test procedure provided by the manufacturer of the test equipment. (See Fig. C.4.)

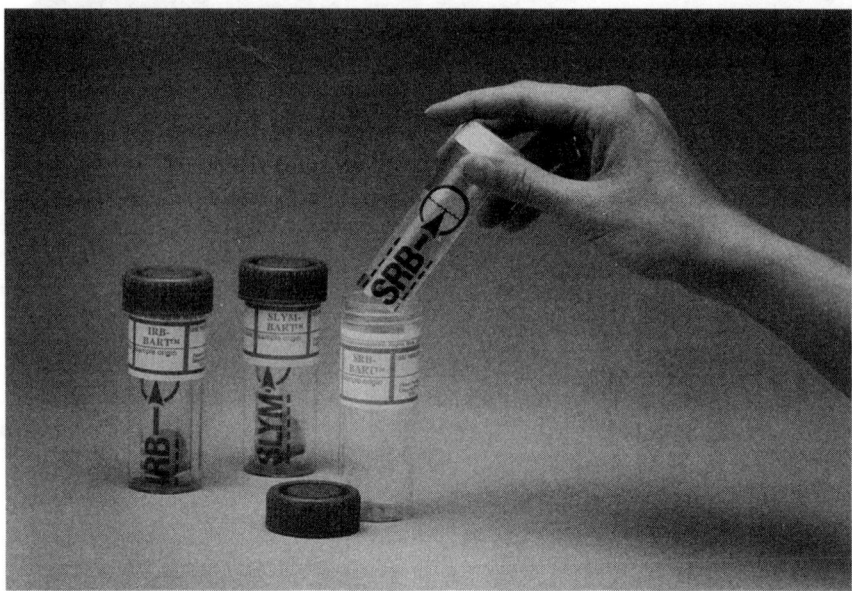

Figure C.4 Biological activity reaction tests for detecting specific algae and bacteria in water. (*Hach Co., P.O. Box 839, Loveland, CO 80539.*)

Appendix D
Computer Programs

Two computer programs, CORSCNEF and WTRCHEM84, are included in this appendix.

By using the CORSCNEF program and knowing the analytical content of the water supply, one can determine the various indexes, namely the Langelier Saturation, Ryznar, Aggressiveness, Calcium Carbonate Precipitation Potential, Momentary Excess, Driving Force, Singley Corrosion, and Larson Indexes. With knowledge of all these indexes, the water chemist is then able to predict more exactly how corrosive or scale-forming a water may be in varied environments and is better able to decide what chemical treatment and metals should be installed to attain economical and efficient operation.

The WTRCHEM84 program supplements the above results by taking into consideration the possible effect of the mixing of two or more waters. Also, the Langelier Saturation, Driving Force, and Calcium Carbonate Precipitation Potential Indexes are calculated, and the amount and kind of chemical required to attain a specified index is also calculated.

While conflicting results may be obtained in some cases, the overall knowledge obtained is helpful in arriving at the best practical solution.

CORSCNEF

```
30 CLS
50 REM CORROSIVITY AND DEPOSITION INDICES "CORSCNEF" PROGRAM
```

Appendix D

```
70 REM C. NEFF, 11/1/87, VERSION 3
100 LPRINT "******************************************"
110 LPRINT
120 LPRINT"           POTABLE WATER INDICES"
130 LPRINT "               DATA ENTRY"
140 LPRINT
150 LPRINT "******************************************"
160 INPUT "SAMPLE ID .............."; ID$
170 LPRINT "SAMPLE NO"; ID$
180 INPUT "MEASURED PH (SITE) ............."; MP
182 LPRINT "MEASURED PH(SITE)"; MP
185 IF MP <4.5 OR MP>11 THEN 180
190 INPUT "TEMPERATURE   (DEG C @ SITE......"; CS
192 LPRINT "TEMPERATURE"; CS
195 IF CS <0 OR CS>100 THEN 190
200 INPUT "CALCIUM   (MG/L AS CACO3)........"; CA
202 LPRINT "CALCIUM (MG/L AS CACO3)"; CA
205 IF CA <= 0 THEN 200
230 INPUT "ALKALINITY   (MG/L AS CACO3)......"; TA
232 LPRINT "ALKALINITY (MG/L AS CACO3)"; TA
235 IF TA <= 0 THEN 230
240 INPUT "DISSOLVED SOLIDS    (MG/L)........"; DS
242 LPRINT "DISSOLVED SOLIDS (MG/L)"; DS
245 INPUT "CHLORIDE (MG/L AS CL)............"; CL
247 LPRINT "CHLORIDE (MG/L AS CL)"; CL
250 INPUT "SULFATE (MG/L AS SO4)............"; SO
252 LPRINT "SULFATE (MG/L AS SO4"; SO
255 INPUT "DISSOLVED OXYGEN    (MG/L DIS.O2).."; O2
257 LPRINT "DISSOLVED OXYGEN (MG/L DIS. O2)"; O2
260 INPUT "REFERENCE TEMP. (DEG.C)........."; CR
262 LPRINT "REFERENCE TEMP. (DEG.C)"; CR
265 IF CR <0 OR CR>100 THEN 260
330 EA = TA / 50045
340 I = DS / 40000: REM IONIC STRENGTH
350 TS = CS + 273.15: REM DEG.KELVIN
355 TR = CR + 273.15
360 DEF FNA (RO) = INT(RO * 10 + .5) / 10: REM FUNCTIONS TO ROUND DATA
370 DEF FNB (RO) = INT(RO * 100 + .5) / 100
380 DEF FNC (RO) = INT(RO * 1000 + .5) / 1000
390 L = 1 / LOG(10)
400 T = TS: C = CS
410 GOSUB 2000: REM EQUILIBRIUM CONSTANTS AT SAMPLE TEMP.
420 H = 10 ^ (-MP - YH): REM H ION CONCENTRATION MOL/L
430 OH = KW / H
450 GOSUB 3000: REM CARBON SPECIATION
460 GOSUB 4000: REM CALCULATE INDICES
470 N = 1
480 GOSUB 7000: REM STORE VALUES COMPUTED FOR SAMPLE TEMPERATURES
500 T = TR: C = CR
510 GOSUB 2000: REM EQUILIBRIUM CONSTANTS AT REFERENCE TEMP.
530 GOSUB 5000: REM COMPUTE PH CHANGE DUE TO TEMP.-NEWTON'S METHOD
540 H = ABS(X)
550 OH = KW / H
560 MP = -LOG(H * 10 ^ YH) * L
565 GOSUB 8000: REM RECALCULATE SPECIATION
570 GOSUB 4000: REM RECALCULATE REFERENCE TEMPERATURE INDICES
580 N = 2
590 GOSUB 7000: REM STORE VALUES COMPUTED FOR REFERENCE TEMPERATURE
600 GOSUB 6150
610 PRINT : PRINT
630 PRINT "              1. - MORE CALCULATIONS "
```

```
640 PRINT "              2. - QUIT"
650 PRINT : PRINT
660 INPUT "              ENTER SELECTION "; Z%
670 IF Z% = 1 THEN 50
680 IF Z% = 2 THEN 760
710 GOSUB 6000
730 GOTO 610
760 END
1000 REM***** ERROR STATEMENTS******
1005 REM*FLAG UNREAL ALKALINITY VALUE *
1020 LPRINT "ERROR IN INPUT!!!-IMPOSSIBLE PH AND"
1030 LPRINT "ALKALINITY COMBINATION, HYDROXIDE      "
1040 LPRINT "EXCEEDS TOTAL ALKALINITY"
1050 LPRINT "HIT ANY KEY TO CONTINUE"
1100 REM * FLAG UNREAL LANGELIER VALUE *
1110 LPRINT "LANGELIER INDEX HAS NO REAL VALUE "
1115 LPRINT "HIT ANY KEY TO CONTINUE"
2000 REM ACTIVITY COEFFICIENTS BY MODIFIED DEBYE HUCKEL
2010 REM TRUESDALL & JONES (1974)
2020 D = -7.047968 + .016796 * T + 1795.711 / T - .0000141566# * T * T
- 153541 / (T * T): REM DENSITY
2030 DW = 87.74 - .4008 * C + .0009398 * C * C - 1.41E-06 * C * C * C:
REM DIELECTRIC CONSTANT
2040 A = 1824600 * (D ^ .5) / (DW * T) ^ 1.5: REM DEBYE A
2050 B = 50.29 * D ^ .5 / (DW * T) ^ .5: REM DEBYE B
2055 SQ = SQR(I)
2060 YC = ((-A * 4 * SQ) / (1 + (B * 5 * SQ))) + .165 * I: REM ACTIV-
ITY COFF. CALCIUM
2070 YR = -A * 4 * SQ / (1 + (B * 5.4 * SQ)): REM ACTIVITY COFF. CAR-
BONATE
2080 YA = -A * 1 * SQ / (1 + (B * 5.4 * SQ)): REM ACTIVITY COFF. BI-
CARBONATE
2090 YH = -A * 1 * SQ / (1 + (B * 9 * SQ)): REM ACTIVITY COFF. HYDRO-
GEN
2100 YX = -A * 1 * SQ / (1 + (B * 3.5 * SQ)): REM ACTIVITY COFF. HY-
DROXIDE
2110 YU = -.5 * I: REM UNCHARGED SPECIES
2130 REM **EQUILIBRIUM CONSTANTS (PK'S)
2140 REM*** PLUMMER & RUSENBERG (1932)*
2150 KC = -171.9065 - .077993 * T + 2839.319 / T + 71.595 * LOG(T) * L:
REM PK CALCITE
2180 KW = 35.3944 - .00853 * T - 5242.39 / T - 11.8261 * LOG(T) * L:
REM WATER
2190 K1 = -356.3094 - .06091964# * T + 21834.37 / T + 126.8339 * LOG(T)
* L - 1684915 / (T * T)
2200 K2 = -107.8871 - .03252849# * T + 5151.79 / T + 38.92561 * LOG(T)
* L - 563713.9 / (T * T)
2210 REM** CONVERT PK VALUES TO K VALUES
2220 REM CORRECT FOR ACTIVITIES
2230 K1 = 10 ^ (K1 + YU - YH - YA): REM K1'
2240 K2 = 10 ^ (K2 + YA - YH - YR): REM K2'
2250 KW = 10 ^ (KW - YH - YX): REM KW'
2260 KC = 10 ^ (KC + YU - YC - YR)
2290 RETURN
3000 REM ******** SPECIATION ********
3010 CD = (H * H) + (K1 * H) + (K1 * K2)
3020 R1 = H * H / CD: REM CO2/IC RATIO
3030 R2 = H * K1 / CD: REM CO2/IC RATIO
3040 R3 = K1 * K2 / CD: REM CO3/IC RATIO
3050 IC = (EA + H - OH) / (R2 + (2 * R3)): REM TOTAL INORGANIC CARBON
MOL/L
```

```
3055 IF IC <0 THEN 1000
3060 DC = IC * R1: REM TOTAL CO2 MOL/L
3070 BC = IC * R2: REM HCO3 MOL/L
3080 CC = IC * R3: REM CO3 MOL/L
3090 AC = (2 * DC + BC + H - OH): REM ACIDITY EQ/L
3200 RETURN
4000 REM***** INDICES SUBROUTINE *****
4010 REM ******LANGELIER INDEX******
4015 CM = CA / 100089
4020 A = 1 - CM * K2 / KC
4025 B = K2 * (2 - (CM * EA / KC))
4030 C = KW * K2 * CM / KC
4035 QQ = B * B - 4 * A * C
4040 IF B * B <4 * A * C THEN 1100
4055 S1 = (-B + SQR(QQ)) / 2 * A
4065 IF S1 <0 THEN 1100
4070 S1 = -LOG(S1 * 10 ^ YH) * L: REM PHS
4075 SC = (MP - S1): REM CALCITE INDEX
4080 REM ********LARSON INDEX ********
4085 LI = (.0282 * CL) + (.0208 * SO) / (.02 * TA)
4090 REM ******** RYZNAR INDEX *******
4095 RI = (2 * S1) - MP
4100 REM****** AGGRESSIVE INDEX*******
4105 AI = MP + LOG(TA * CA) * L
4110 REM ****** SNOEYINK INDEX *******
4115 HS = CM * (EA - (KW / H) + H) * (K2 / KC) * (H / ((2 * K2) + H)):
HS = ABS(HS)
4120 S2 = -LOG(HS * 10 ^ YH) * L: REM PHS
4125 SS = MP - S2: REM CALCITE INDEX
4130 REM ******DRIVING FORCE INDEX******
4135 FI = (CM * CC) / KC
4140 REM ***MOMENTARY EXCESS INDEX ***
4145 Q2 = -CM - (EA + H - OH) * (K2 / (H + (2 * K2)))
4150 Q3 = -KC + (CM * (EA + H - (KW / H)) * (K2 / (H + (2 * K2))))
4155 ME = (-Q2 - SQR(Q2 * Q2 - 4 * Q3)) / 2
4160 ME = ME * 100089
4165 REM ****** BUFFER CAPACITY ******
4170 B1 = R3 * H * ((EA * H) + (H * H) - KW)
4175 B2 = (K1 * H) + (2 * K1 * K2)
4180 B3 = (H / K2) + (K1 / H) + 4
4185 BI = 2.303 * ((B1 * B3 / B2) + OH + H)
4190 BI = BI * 50045
4195 REM ******* SINGLEY INDEX *******
4200 JS = (CL ^ .509) * (SO ^ .0249) * (TA ^ .423) * (O2 ^ .78)
4205 RP = ((.4 * CA) ^ .676) * (BI ^ .0304) * ((10 ^ SC) ^ .107) * 9.4675
4210 SG = JS / RP: REM SINGLEY & PISIGAN, MATERIALS PERFORMANCE APRIL
1985
4215 REM SG=CORROSION RATE OF MILD STEEL (mpy)
4240 REM *******CCPP INDEX *******
4250 HE = 10 ^ (-S1 - YH)
4255 TE = ((2 * K2) + HE) / HE
4260 SE = HE - (KW / HE)
4265 PE = ((2 * HE) + K1) / K1
4270 EC = ((TE / PE) * (AC - SE)) - SE
4275 EC = ABS(EC)
4295 PP = EA - EC
4299 RETURN
5000 REM** COMPUTE HYDROGEN ION AT NEW TEMPERATURE FROM TOTAL ACIDITY
VALUE **
5010 REM ***** USING NEWTON'S METHOD *****
5020 V1 = 1
```

```
5030 V2 = K1 + EA
5040 V3 = -((KW - K1 * K2) + K1 * (IC - EA))
5050 V4 = -K1 * (KW + (K2 * ((2 * IC) - EA)))
5060 V5 = -K1 * K2 * KW
5070 IF MP>7 THEN EP = MP - 1.8: REM ESTIMATE FOR NEWTON'S METHOD
5080 IF MP <7 THEN EP = MP - .15
5090 IF MP = 7 THEN EP = 7
5100 X = 10 ^ (-EP)
5110 TL = .000001
5115 R = 1E-15
5120 FOR W = 1 TO 30
5130 X1 = X
5140 GOSUB 5500
5145 IF (ABS(DF)>R) THEN 5160
5160 DX = FX / DF
5170 X = X1 - DX
5180 IF (ABS(DX) <= ABS(TL * X)) THEN 5210
5190 NEXT W
5195 LPRINT : LPRINT
5200 LPRINT "ERROR - NO CONVERGENCE IF 30 ITERATIONS !!"
5210 RETURN
5500 FX = X ^ 4 + V2 * X ^ 3 + V3 * X ^ 2 + V4 * X + V5
5510 DF = 4 * X ^ 3 + 3 * V2 * X ^ 2 + 2 * V3 * X + V4
5520 RETURN
6000 REM********** DATA OUTPUT *********
6020 LPRINT
6050 LPRINT "SAMPLE IDENTIFICATION   "; ID$
6060 LPRINT
6070 LPRINT "MEASURED PH (ST.UNITS)  "; MP(1)
6080 LPRINT "TEMPERATURE (DEG.CELSIUS)"; CS
6090 LPRINT "ALKALINITY (MG/L CACO3)  "; TA
6100 LPRINT "CALCIUM     (MG/L CACO3)   "; CA
6110 LPRINT "TDS         (MG/L)         "; DS
6120 LPRINT "CHLORIDE    (MG/L CHLORIDE)"; CL
6130 LPRINT "SULFATE     (MG/L SULFATE) "; SO
6135 LPRINT "DIS.OXYGEN (MG/L OXYGEN)   "; O2
6140 LPRINT "REFERENCE TEMP. (DEG.C)    "; CR
6150 LPRINT
6160 LPRINT "SPECIATION OF CO2 - MG/L AS CACO3"
6165 LPRINT "         ("; CS; " DEG.CELSIUS)"
6170 LPRINT
6180 LPRINT "TOTAL ACIDITY            "; AC(1)
6200 LPRINT "TOTAL DIS. CO2           "; DC(1)
6210 LPRINT "BICARBONATE              "; BC(1)
6220 LPRINT "CARBONATE                "; CC(1)
6230 LPRINT "HYDROXIDE                "; OH(1)
6240 LPRINT
6260 LPRINT
6265 LPRINT "INDICES FOR "; CS; "DEG. CELSIUS"
6270 LPRINT
6280 LPRINT "SATURATION PH (LANGELIER)  "; PH(1)
6285 LPRINT "SATURATION PH (SNOEYINK)   "; PS(1)
6290 LPRINT "LANGELIER INDEX            "; SC(1)
6295 LPRINT "SNOEYINK INDEX             "; SS(1)
6300 LPRINT "RYZNAR INDEX               "; RI(1)
6310 LPRINT "AGGRESSIVE INDEX           "; AI(1)
6320 LPRINT "CCPP (MG/L CACO3)          "; PP(1)
6330 LPRINT "MOMENTARYEXCESS(MG/L CACVO3)"; ME(1)
6340 LPRINT "BUFFERCAPACITY (MG/L CACO3) "; BI(1)
6345 LPRINT "DRIV.FORCEINDEX            "; FI(1)
6350 LPRINT "SINGLEYCORR.RATE (mpy)     "; SG(1)
```

268 Appendix D

```
6360 LPRINT "LARSON INDEX                 "; LI(1)
6400 LPRINT
6410 LPRINT " SPECIATION OF CO2 -MG/L AS CACO3"
6420 LPRINT "         ("; CR; " DEG.CELSIUS)"
6430 LPRINT
6440 LPRINT "TOTAL ACIDITY              "; AC(2)
6450 LPRINT "TOTAL DISSOLVED CO2        "; DC(2)
6460 LPRINT "BICARBONATE                "; BC(2)
6470 LPRINT "CARBONATE                  "; CC(2)
6475 LPRINT "HYDROXIDE                  "; OH(2)
6480 LPRINT
6500 LPRINT
6505 LPRINT " INDICES FOR "; CR; " DEG. CELSIUS"
6510 LPRINT "(ASSUMES NO LOSS OR GAIN OF CO2)"
6520 LPRINT
6530 LPRINT "ESTIMATED PH @ DEG. C      "; MP(2)
6540 LPRINT "PH CHANGE DUE TO TEMP.     "; DP
6550 LPRINT "SATURATION PH @ DEG. C     "; PH(2)
6560 LPRINT "LANGELIER INDEX            "; SC(2)
6565 LPRINT "SNOEYINK INDEX             "; SS(2)
6570 LPRINT "RYZNAR INDEX               "; RI(2)
6580 LPRINT "AGGRESSIVE INDEX           "; AI(2)
6590 LPRINT "CCPP (MG/L CACO3)          "; PP(2)
6600 LPRINT "MOMENTARY EXCESS (MG/l CACO3) "; ME(2)
6610 LPRINT "BUFFER CAPACITY (MG/L CACO3) "; BI(2)
6620 LPRINT "DRIVINGFORCE INDEX (MG/L CACO3) "; FI(2)
6630 LPRINT "SINGLEY CORR. RATE (mpy)   "; SG(2)
6650 RETURN
7000 REM****ARRAY FOR TEMPERATURE****
7010 MP(N) = FNB(MP): OH(N) = FNB(OH * 50045)
7015 PH(N) = FNB(S1): PS(N) = FNB(S2): SC(N) = FNB(SC): SS(N) = FNB(SS)
7020 RI(N) = FNA(RI): AI(N) = FNA(AI): LI(N) = FNB(LI): BI(N) = FNB(BI)
7030 FI(N) = FNA(FI): ME(N) = FNB(ME): SG(N) = FNB(SG): PP(N) = FNB(PP
* 50045)
7035 DC(N) = FNA(DC * 100089): BC(N) = FNA(BC * 50045): CC(N) = FNA(CC
* 100089)
7050 AC(N) = FNA(AC * 50045)
7055 DP = MP(2) - MP(1): DP = FNB(DP)
7065 RETURN
8000 REM RECALCULATE SPECIATION
8010 CD = (H * H) + (K1 * H) + (K1 * K2)
8020 R1 = H * H / CD: REM CO2/IC RATIO
8030 R2 = H * K1 / CD: REM HCO3/IC RATIO
8040 R3 = K1 * K2 / CD: REM CO3/IC RATIO
8060 DC = IC * R1: REM TOTAL CO2 MOL/L
8070 BC = IC * R2: REM HCO3 MOL/L
8080 CC = IC * R3: REM CO3 MOL/L
8090 AC = (2 * DC + BC + H - OH): REM ACIDITY EQ/L
8110 RETURN
```

WTRCHEM84

```
100 REM WTRCHEM-JRR 030784
120 REM USING MERRILL'S ALGORITHM
125 REM    LPRINT
130 T5 = 273.16
140 L1 = 1 / LOG(10)
150 E1 = 50000
160 E2 = 100000
```

```
170 PRINT "HOW MANY SUPPLIES TO BE MIXED";
180 INPUT N
190 FOR Y1 = 1 TO N
200 IF N - Y1 = 0 THEN 260
210 PRINT "PERCENT FROM SUPPLY#"; Y1; "     ";
220 INPUT F1: F1 = F1 / 100
230 IF F1 <= 0 OR F1 >= 1 THEN 280
240 F2 = F2 + F1
250 GOTO 320
260 F1 = 1 - F2
270 IF F1>0 AND F1 <= 1 THEN 300
280 PRINT "THERE'S AN ERROR. PERCENT MUST BE BETWEEN 0 AND 100."
290 STOP
300 PRINT
310 PRINT "FOR SUPPLY #"; N
320 PRINT "TEMPERATURE (DEG C)          ";
330 INPUT T1
340 T2 = T2 + T1 * F1
350 T3 = 25
360 PRINT "DISSOLVED SOLIDS (MG/L)      ";
370 INPUT U1
380 U1 = U1 / 40000
390 U2 = U2 + U1 * F1
400 PRINT "ALKALINITY MG/L AS CACO3)       ";
410 INPUT A1
420 A1 = A1 / E1
430 A2 = A2 + A1 * F1
440 PRINT "CALCIUM (AS CACO3)           ";
450 INPUT C1
460 C1 = C1 / E2
470 C2 = C2 + F1 * C1
480 PRINT "MAGNESIUM (AS CACO3)          ";
490 INPUT M1
500 M1 = M1 / E2
510 M2 = M2 + F1 * M1
520 PRINT "PH @"; T3; "DEG. C              ";
530 INPUT H1
540 GOSUB 3000
550 H = 10 ^ (M - H1)
560 P = (1 + 2 * H / K1) / (1 + 2 * K2 / H)
570 q = H - KW / H
580 D1 = A1 * P + P * q + q
590 D2 = D2 + F1 * D1
600 NEXT Y1
610 PRINT
620 PRINT "ANALYSIS @"; T2; "C"
630 T3 = T2
640 U1 = U2
650 C1 = C2
660 A1 = A2
670 M1 = M2
680 D1 = D2
690 GOSUB 3000
700 IF A1 + q>0 THEN 730
710 PRINT "THERE'S AN ERROR IN DATA. ALKY MUST BE MORE THAN HYDROXYL."
720 STOP
730 H = .001
740 D = 10
750 H = H / (D + 1)
760 q = H - KW / H
770 P = (1 + 2 * H / K1) / (1 + 2 * K2 / H)
```

```
780 D2 = A1 * P + P * q + q
790 IF D1 <D2 THEN 750
800 H = H * (1 + D)
810 D = D / 2
820 IF D>.000001 THEN 750
830 H1 = M - L1 * LOG(H)
840 B1 = (A1 + q) / (1 + 2 * K2 / H)
850 G1 = H * B1 / K1
860 J1 = K2 * B1 / H
870 KS = 10 ^ (13.87 - 3059 / T4 - .04035 * T4 + 8 * M)
880 KM = 10 ^ (6 * M - 9.97 - .0175 * T3)
890 A = 1 - K2 * C1 / KS
900 B = K2 * (2 - C1 * A1 / KS)
910 C = K2 * KW * C1 / KS
920 IF B ^ 2>4 * A * C THEN 960
930 L$ = "NO REAL VALUE"
940 S = -1
950 GOTO 1020
960 HS = (SQR(B ^ 2 - 4 * A * C) - B) / 2 / A
970 IF HS <= 0 THEN 930
980 H2 = M - L1 * LOG(HS)
990 S1 = (H1 - H2)
1000 L$ = STR$(INT(1000 * (S1) + .5) / 1000)
1010 S = SGN(S1)
1020 PRINT
1030 PRINT TAB(5); "CALCIUM"; TAB(25); INT(100 * C1 * E2 + .5) / 100
1040 PRINT TAB(5); "MAGNESIUM"; TAB(25); INT(100 * M1 * E2 + .5) / 100
1050 PRINT TAB(5); "PH"; TAB(25); INT(100 * H1 + .5) / 100
1060 PRINT TAB(5); "ALKALINITY"; TAB(25); INT(100 * A1 * E1 + .5) / 100
1070 PRINT TAB(5); "BICARBONATE"; TAB(25); INT(100 * B1 * E1 + .5) / 100
1080 PRINT TAB(5); "CARBONATE"; TAB(25); INT(100 * J1 * E1 + .5) / 100
1090 PRINT TAB(5); "CARBON DIOXIDE"; TAB(25); INT(100 * G1 * E2 + .5) / 100
1100 PRINT TAB(5); "ACIDITY"; TAB(25); INT(100 * D1 * E1 + .5) / 100
1110 PRINT TAB(5); "LANGELIER INDEX"; TAB(25); L$
1120 Z1 = J1 * C1 / KS
1130 PRINT TAB(5); "DRIVING FORCE INDEX"; INT(1000 * Z1 + .5) / 1000
1140 D = S / 2
1150 H = H / (1 - D)
1160 Z = M1 - KM * (H / KW) ^ 2
1170 IF Z <0 THEN Z = 0
1180 M2 = M1 - Z
1190 A2 = A1 - 2 * Z
1200 D1 = D2 + 2 * Z
1210 P = (1 + 2 * H / K1) / (1 + 2 * K2 / H)
1220 q = H - KW / H
1230 R = 2 + H / K2
1240 IF D1 - q <= 0 THEN 1270
1250 Q8 = 2 * KS * R * P / (D1 - q) - (D1 - q) / P + q - 2 * C1 + A2
1260 IF S * Q8 <0 THEN 1150
1270 H = H * (1 - D)
1280 D = D / 2
1290 IF ABS(D)>.000001 THEN 1150
1300 A3 = (D1 - q) / P - q
1310 X = (A2 - A3) / 2
1320 M1 = M2
1330 A1 = A2
1340 PRINT
1350 PRINT "EQUILIBRATED STATE:"
1360 PRINT
```

```
1370 PRINT "      PH"; TAB(25); INT(100 * (M - L1 * LOG(H)) + .5) / 100
1380 PRINT "      CALCIUM"; TAB(25); INT(100 * E2 * (C1 - X) + .5) / 100
1390 PRINT "      MAGNESIUM"; TAB(25); INT(100 * E2 * M1 + .5) / 100
1400 PRINT "      ALKALINITY"; TAB(25); INT(100 * E1 * (A1 - 2 * X) +
.5) / 100
1410 PRINT " PRECIPITATION POTENTIALS:"
1420 PRINT "      CACO3"; TAB(25); INT(100 * E2 * X + .5) / 100
1430 PRINT "      MG(OH)2"; TAB(25); INT(100 * E2 * Z + .5) / 100
1440 PRINT
1450 IF I = 1 THEN 1480
1460 PRINT "IS TREATMENT DESIRED";
1470 GOTO 1499
1480 PRINT "IS FURTHER TREATMENT DESIRED";
1495 PRINT : PRINT
1496 PRINT "           1. -more calculations"
1497 PRINT "           2. -quit"
1498 PRINT : PRINT
1499 INPUT q$
1500 IF q$ = "1" THEN 130
1501 IF q$ = "2" THEN 2710
1510 PRINT
1520 PRINT "SELECT TREATMENT:"
1530 PRINT "      1.CALCIUM        5.LIME"
1540 PRINT "      2. MAGNESIUM     6.SODA ASH"
1550 PRINT "      3. ACID          7.CARBON DIOXIDE"
1560 PRINT "      4. CAUSTIC       8. BICARBONATE"
1570 PRINT "        9. TREAT TO A SPECIFIED CONDITION"
1580 PRINT ; TAB(7); "0. TERMINATES SELECTIONS"
1590 I = 1: PRINT
1610 PRINT "SELECTION";
1620 INPUT V: PRINT
1630 IF V = 0 THEN 2030
1650 ON V GOTO 1660, 1700, 1740, 1790, 1840, 1900, 1940, 1980, 2050
1660 PRINT "CALCIUM DOSE";
1670 INPUT D
1680 C1 = C1 + D / E2
1690 GOTO 1610
1700 PRINT "MAGNESIUM DOSE";
1710 INPUT D
1720 M1 = M1 + D / E2
1730 GOTO 1610
1740 PRINT "ACID DOSE";
1750 INPUT D
1760 A1 = A1 - D / E1
1770 D1 = D1 + D / E1
1780 GOTO 1610
1790 PRINT "CAUSTIC DOSE";
1800 INPUT D
1810 A1 = A1 + D / E1
1820 D1 = D1 - D / E1
1830 GOTO 1610
1840 LPRINT "LIME DOSE";
1850 INPUT D
1860 C1 = C1 + D / E2
1870 A1 = A1 + D / E1
1880 D1 = D1 - D / E1
1890 GOTO 1610
1900 PRINT "SODA ASH DOSE";
1910 INPUT D
1920 A1 = A1 + D / E1
1930 GOTO 1610
```

Appendix D

```
1940 PRINT "CARBON DIOXIDE DOSE";
1950 INPUT D
1960 D1 = D1 + D / E1
1970 GOTO 1610
1980 PRINT "BICARBONATE DOSE";
1990 INPUT D
2000 D1 = D1 + D / E1
2010 A1 = A1 + D / E1
2020 GOTO 1610
2030 PRINT "INTERIM STATE";
2040 GOTO 730
2050 PRINT "TREAT TO A SPECIFIED VALUE OF:"
2060 PRINT
2070 PRINT " 1.LANGELIER INDEX"
2080 PRINT "2.DRIVING FORCE INDEX"
2090 PRINT "3.CCPP"
2100 PRINT
2110 PRINT "SELECTION"
2120 L = 1
2130 INPUT S
2140 PRINT
2150 ON S GOTO 2160, 2210, 2260
2160 PRINT "LANGELIER INDEX DESIRED";
2170 INPUT I
2180 L = 10 ^ I
2190 S = SGN(INT(100 * (S1 - I) + .5))
2200 GOTO 2300
2210 PRINT "DRIVING FORCE DESIRED";
2220 INPUT Z
2230 KS = KS * Z
2240 S = SGN(INT(100 * (1 - Z / Z1) + .5))
2250 GOTO 2300
2260 PRINT "VALUE OF CCPP DESIRED";
2270 INPUT X1
2280 S = SGN(INT(E2 * X - X1) + .5)
2290 X1 = X1 / E2
2300 PRINT
2310 PRINT "SELECT TREATMENT"
2320 ON S + 2 GOTO 2330, 2700, 2360
2330 PRINT "1.LIME        2.CAUSTIC         3.SODA ASH";
2340 INPUT V
2350 GOTO 2390
2360 PRINT "1.ACID      2. CARBON DIOXIDE";
2370 INPUT V1
2380 V = V1 + 3
2390 PRINT
2400 I = 1
2410 D = .001 * S
2420 C2 = C1 - X1
2430 A2 = A2 - 2 * X1
2440 Y = Y - D
2450 ON V GOTO 2460, 2470, 2490, 2470, 2470
2460 C2 = C1 + Y - X1
2470 D2 = D1 - 2 * Y
2480 IF V = 5 THEN 2500
2490 A2 = A1 + 2 * Y - 2 * X1
2500 A = 1 - K2 * C2 / KS
2510 B = K2 * (2 - C2 * A2 / KS)
2520 C = K2 * KW * C2 / KS
2530 IF 4 * A * C>B ^ 2 THEN 2440
2540 H = (SQR(B ^ 2 - 4 * A * C) - B) / 2 * A / L
```

```
2550 q = H - KW / H
2560 P = (1 + 2 * H / K1) / (1 + 2 * K2 / H)
2570 D3 = A2 * P + P * q + q
2580 IF S * D3>S * D2 THEN 2440
2600 Y = Y + D
2610 D = D / 2
2620 IF ABS(D)>1E-08 THEN 2440
2630 PRINT "DOSE REQUIRED:"; INT((-S * 100 * E2 * Y + .5) / 100)
2640 ON V GOTO 2650, 2660, 2680, 2660, 2660
2650 C1 = C1 + Y
2660 D1 = D1 - 2 * Y
2670 IF V = 5 THEN 2690
2680 A1 = A1 + 2 * Y
2690 GOTO 2030
2700 PRINT "NO TREATMENT REQUIRED"
2702 INPUT "want to calculate more data (Y/N)"; A$
2704 IF A$ = "Y" THEN 130
2706 PRINT
2710 END
3000 E = 60954 / (T3 + 389) - 68.937
3010 T4 = T3 + T5
3020 A = 1825000 / (E * T4) ^ 1.5
3030 M = A * (SQR(U1) / (1 + SQR(U1)) - .2 * U1)
3040 K1 = 10 ^ (14.8435 - 3404.71 / T4 - .032786 * T4 + 2 * M)
3050 K2 = 10 ^ (6.498 - 2902.39 / T4 - .02379 * T4 + 4 * M)
3060 KW = 10 ^ (6.0846 - 4471.33 / T4 - .017053 * T4 + 2 * M)
3070 RETURN
```

Index

Air-conditioning systems, 111
Air washers, 138
Algae, 137–139
Alkalinity of water, 6, 243–247
  low alkalinity water, 60
Ambient (free) cooling, 112–113, 155
Amines, 93–95
Analytical methods, 239–262
Anodes, cathodes role in corrosion process, 12, 15
Antifreeze, use in closed recirculating systems, 156, 157
ASTM D2688 corrosion test, 21

Bacteria, total count, 262
  sulfate-reducing bacteria, 142, 262
Biological growths, 138–143
Boiler(s), 70
  ASME Boiler and Pressure Vessel Code, 70
  electric, 100, 102
  hot water, steam, and heating, 69, 78, 162
  operational techniques for improving efficiency, 105, 106
  types of high-pressure steam boilers, 96–100
Boiler feedwater treatment, 70–72
  benefits, 72
  chemical feed systems, 86, 182–183
  continuous blowdown, 87, 89
  control chart, 88
  control of scale formation, 72
  dealkalization, 92
  demineralization, 92
  humidification, 79
  internal water quality limits for 0–300 psig steam boilers, 75

Boiler feedwater treatment (*Cont.*):
  ion exchange, 76–78
  Larson-Lane Steam Purity Analyzer, Condensate Reboiler, 105
  need for treating whole system, 71
  objectives, 72
  postchemical treatment, 78
  prevention of carryover, 83
  prevention of corrosion, 79
  prevention of stress corrosion cracking, 82
  proper sampling, 84, 85, 238
  steam purity limits and control, 101, 103–105
  treatment for condensate return systems, 93–95
  water quality requirements for high-pressure boilers, 73
  water test limits, 90–91
Boiler plant operations, 70, 95–98
  boiling out new boilers, 106
  deaerator, 80–82
  lay-up of out-of-service boilers, 107–108
  makeup percentage, 70
Bottled water, 49
Building systems and maintenance, 197
  backflow prevention, 202
  drinking water systems, 202
  fire protection systems, 203
  materials, 198, 199
  monitoring, 199, 200
  nitrogen-blanketed water tanks, 202
  zones, 197
Buffer intensity (buffer capacity), 59

Cadmium in water, 67

Case histories, 40
Cast iron, 186
Cathodic protection, 19, 30, 39, 143
Causes, common, of corrosion and scale problems, 8
Chelants, 102, 261
Chemical feed systems, 182, 183
Chlorination, 47, 48
  chloramine, 47, 48
  trihalomethanes, 47
Clarification of water, 171, 172
Closed system methods of water treatment, 159–161
  chromate, 159
  molybdate–azole, 161
  nitrite-azole, 160
  oxygen scavenger type, 160
  silicate, 161
Closed systems (CW, MTW, HTW), 149
  cleaning of systems, 162–164
  corrosion and its control, 150
  definition of a closed system, 149
  fire protection systems 164, 165
  fouling and its control, 151
  glycol systems, 156, 157
  hot water closed systems, 158
  maintenance of leak-free systems and test, 150, 152–154
  thermal energy systems, 155, 203
  types of closed systems, chilled and hot, 149
  water losses from and tests for, 150, 152–154
Colorimetric analytical methods, 260–262
Computer programs for determination of water indexes and chemical requirements for water treatment, 263–273
Concentration-cell corrosion, 16
Condensate systems, 79, 93–95
  amines, 93–95
  treatment for, and control, 93–95
Conductivity, electrical, 256–260
Contamination, metallic (*see* Metallic contamination)
Cooling tower(s), 111–116
  free (ambient) cooling, 112, 113, 155
  galvanized, 115
  materials, 116
  water treatment and control, 117–121, 155

Cooling tower water treatment and control, 117–121, 155
  calcium sulfate scale control, 127
  calculation of treatment and blowdown, 117
  chemical feed systems, 119, 182–183
  control of fouling:
    algae, slime, 137–139
    nonoxidizing biocides, 140, 141
    oxidizing biocides, 139–141
  corrosion control and inhibitor information, 129, 131
  corrosion coupon testing, 132, 133
  general treatment programs (chromate, molybdate, azole, zinc), 130, 135–137
  Legionnaire's disease, 143–145
  objectives of cooling water treatment, 122
  sampling, 237
  scale control, 122–129
  tests, precautions to avoid fouling, 142
Cooling water systems, 111
  components, 134
  lay-up of out-of-service equipment, 146, 147
  open recirculating cooling water systems, 111
  pretreatment of cooling water equipment, 145, 146
  role of cooling towers, 111, 112
  sampling, 237
Copper-bearing alloys, 187, 188
Copper corrosion, 31
Corrosion, 11–42
  of brass, 39
  causes, 8
  components, necessary for corrosion to proceed, 12
  concentration-cell, 16
  of copper in water, 31–34
    methods for inhibiting, 37
    pitting, 35
  corrosion rate testing, 12, 13, 21, 52, 131–133, 162
  cost of, 2, 11
  crevice, 16
  definition, 2
  design to avoid corrosion problems, 2
  dezincification, 17
  electrochemical process, 11
  equations of corrosion reactions, 14–15

Corrosion (*Cont.*):
  erosion-corrosion, 18
  evidence of corrosion, 2, 11
  galvanic, 16, 29
  of galvanized steel in water, 25–27
    effect of soluble copper, 29, 53
  graphitization, 17
  how to correct corrosion problems, 8
  pitting, 16
  of pumps, 193
  stray current, 18
  stress corrosion, 17
  tests, simple for diagnosis, 3, 179
  types of, 16
  uniform (general), 16
  of valves, 192
Corrosion inhibitors: anodic, cathodic and mixed, 19
Crevice corrosion, 16
Cycles of concentration (COC), 118
Cyclohexylamine, 93–95

Deaeration, 79–81
  ensuring proper functioning of deaerator, 80–81
Dealkalization, 179–180
Deionization, 178
Dielectric unions, 19
Dissolved oxygen, solubility, 83
Dissolved solids (conductivity test), 256–260
Domestic hot water systems, 197
  city codes, 198
  corrosion and scale problems in, 50–52
  deposit appearances, 4
  design of new and redesign of old, 2
  need for water treatment, 2
  temperature to be maintained, 190

Electric grounding, 31
Electrodialysis, 179
Erosion-corrosion, 18

Federal drinking water standards, 44
Feedwater treatment (*see* Boiler feedwater treatment)
Filtration, 168
  nanofiltration, 179
  ultrafiltration, 179
Fire protection systems, 164, 203
Flux, as a corrodent, 37
Free (ambient) cooling, 112, 113, 155

Gadgets, 181
Galvanic corrosion, 16, 29
  effect of soluble copper, 29
Grains per gallon, 7

Hardness, 5, 6, 240–243
Heat exchangers, 69
High-purity water, methods producing, 175
Hot water boiler water treatment, 162
Humidification, 79

Ice makers, 46
Indexes for estimating corrosion and scaling tendencies, 3
  Langelier, 47, 53–56, 171, 186, 200
  Larson, 47, 59
  Other (such as Driving Force, Momentary Excess, Aggressiveness, Calcium Carbonate Precipitation Potential, Riddick's), 58–59
  Ryznar, 56–58
  Singley, 56
Instantaneous heaters, 4
Ion exchange, 171, 173, 178–181
Iron and manganese removal, 180

Langelier Saturation Index, 47, 53–56, 171, 186, 200
Larson Index, 47, 59
Larson-Lane Steam Purity Analyzer, Condensate Reboiler model, 105
Lead and Copper Rule, 61–63
Lead contamination, 45
Legionnaire's disease, 143–145
Lime-soda softening, 171–173

Materials in water–using systems, 185, 198
Maximum contaminant level (MCL), 44, 160, 163, 166
  aluminum, 66
  copper, 38, 61
  lead, 45, 61, 63
  nitrate, 67
  sodium, 66
Membranes, 176
Metallic contamination, 63, 65
  associated health problems, 43
  copper, 38, 66
  corrosion inhibitor treatment, 49
  lead, 64–66

Metallic contamination (*Cont.*):
  methods of reducing metallic contamination, 65
  revelance of sampling methods, 64
  zinc, 66
Methods of producing high-purity water, 175
  deionization, 178
  electrodialysis, 179
  membranes, 176
  nanofiltration, 179
  reverse osmosis, 178
  ultrafiltration, 179
Methods of water analysis, 239
  alkalinity, 243–246
  chloride, 248, 250
  chromate, 254–256
  field-type colorimetric determinations of molybdate, zinc, and phosphonate, 262
  field-type tests for total bacterial count and sulfate-reducing bacteria, 262
  hardness, calcium, magnesium, 240–243
  instrumental test methods, 256–262
  nitrite, 253–254
  pH, 247–248, 260
  pH estimation from P, M alkalinity, 248, 249
  phosphate, 252–253
  photometric methods, 260–261
  sulfite, 251–252
Microbiological contamination and MIC corrosion, 130, 139, 142
Monitoring building water systems variables, 199
Morpholine, 93–95
Myers' corrosion rate equations, 27, 35

Nanofiltration, 179
National Pollutant Discharge Elimination System (NPDES), 141, 201
Nitrate, 67, 175
  contamination from water, 175

Once-through cooling, 111

Phosphonates, 123, 131
Pitting corrosion, 16
Plastic piping, 189–191
  ABS (acrylonitrile-butadiene-styrene), 190

Plastic piping (*Cont.*):
  CPVC (chlorinated polyvinyl chloride), 190
  PB (polybutylene), 191
  PE (polyethylene), 191
  PP (polypropylene), 190
  PVC (polyvinyl chloride), 190
Point-of-use (POU) treatment, 48, 175
Polyacrylates, 131
Polyphosphates, 29
Pot feeder, 98
Potable water, 43
  central chilled, 46
  corrosion inhibitors for, 49
  criteria for, 43
  federal drinking water standards, 44
  off-flavors, 3
  scale control, 2
POU (point-of-use) treatment, 48, 175
Protective coatings, 188, 189
Puckorius Index, 124
Pump mechanical seals, 194

Quality of work for piping materials, 186

Refrigeration, ton of, 111
Reverse osmosis, 178
Ryznar Index, 56

Safe Drinking Water Act, 43
Sampling, 237
  boiler water, 238
  containers, 237
  from different sources, 237
  relevance of methods with respect to metallic contamination 64–65
  return condensate, 238
Scale, 3
  composition of, 4
  energy loss due to boiler scale, 71
  how to correct problem of, 8
  methods of inhibiting different types of scale, 33
Selection of materials, 185, 198
Silicates as corrosion inhibitors, 51
Singley Index, 56
Slime growths, 137–139
Sodium, 67
Sodium zeolite softening, 75–77, 173–174
Specifications for water-testing reagents and equipment, 225

Specifications for water treatment chemicals, 205
"Stable" water, 171
Stainless steels, for piping, 187, 199
  compared with fiberglass, 188
  deficiencies (such as stress corrosion cracking), 187
Steam purity, determination of, 104
Steam separators, 103
Steel, 186
  (*See also* Stainless steels, for piping)
Stray current corrosion, 18
Stress corrosion, 17
Stress corrosion cracking, 82
Sulfate-reducing bacteria (SRB), 142, 262
Survey of Chicago high-rise buildings, 197, 200, 202–203
  chilled water tanks, 202
  drinking water systems, backflow prevention in, 202
  fire protection systems, 203
  older vs. newly constructed, 200
  water treatment monitoring, 201
Surface water, 6

Test coupons, 7
Tests to determine deposit problem and derive solution, 3
Thermal energy systems, 155
Titration analytical methods, 240–243
Ton of refrigeration, 111
Trihalomethanes, 47

Ultrafiltration, 179
Uniform (general) corrosion, 16

Water, 5–10
  alkalinity of, 6, 243–247
  clarification of, 171, 172
  composition of, 5, 6
  composition in U.S. cities, 6
  constituents important in corrosion process, 20

Water (Cont.):
  expressing concentration, 7
  hardness, 5, 6
  impurities in, 5
  monitoring/testing, 74, 81, 88, 91, 225, 237
  potable (*see* Potable water)
  properties, 5
  uses in buildings, 1
Water analysis methods (*see* Methods of water analysis)
Water-caused problems, 3, 8
  how to correct, 8
  simple tests for identifying, 3
Water quality, 167
  common impurities in water, 169
  different types, 167
  effects on corrosion rate, 2
  effects of water treatment methods on water quality, 168, 171
  municipal water supplies, 167, 168
Water service piping materials, 185, 198
  designing for correct materials, 186
  quality of work, 186
Water testing (*see* Methods of water analysis)
Water treatment:
  effects on corrosion, scaling tendencies, 173
  equipment, need for, 1
  specifications, 205
  (*See also* Closed system methods of water treatment; Cooling tower water treatment and control; Hot water boiler water treatment; Point-of-use (POU) treatment)
Well water, 167

Zeolite softening, 173
  malfunctions, 173, 174
  needed maintenance, 174
Zinc in water, 66, 131

ABOUT THE AUTHOR

Russell W. Lane, P.E., is a water treatment consultant working with industry, universities, institutions, and municipalities. He served with the Illinois State Water Survey for more than 30 years as both a chemist and Head of the Chemistry Section, and now as Principal Scientist Emeritus. Mr. Lane is the author of more than 50 papers on water treatment and testing methods and holds seven patents in the field. He is the recipient of numerous awards for his distinguished work, including the Max Hecht Award for Outstanding Service in the Study of Water, the Award of Merit from the American Society for Testing and Materials (ASTM), and Citation of Recognizance from the National Association of Corrosion Engineers. He is a registered professional engineer in California and Illinois, and the Initiator of the annual Electric Utility Chemistry Workshop at the University of Illinois. Mr. Lane resides in Champaign, Illinois.